CONTROLLING HOUSEHOLD PESTS

Created and designed by
the editorial staff of ORTHO BOOKS

WRITER
Wayne S. Moore

PROJECT EDITOR
Susan A. Roth

ASSOCIATE AND PHOTO EDITOR
Pamela Peirce

ARTIST
Amy Bartlett Wright

PHOTOGRAPHER
Saxon Holt

GRAPHIC DESIGN
Finger Vesik Smith

Ortho Books

Publisher
Robert J. Dolezal

Editorial Director
Christine Robertson

Production Director
Ernie S. Tasaki

Managing Editors
Michael D. Smith
Sally W. Smith

System Manager
Katherine Parker

National Sales Manager
Charles H. Aydelotte

Marketing Specialist
Susan B. Boyle

Circulation Manager
Barbara Steadham

Operations Coordinator
Georgiann Wright

Senior Technical Analyst
J. A. Crozier, Jr.

Chevron Chemical Company
6001 Bollinger Canyon Road
San Ramon, CA 94583

Acknowledgments

Art Director
Craig Bergquist

Copy Chief
Melinda E. Levine

Copyeditor
Rebecca Pepper

Layout & Pagination by
Linda M. Bouchard

Editorial Assistant
Andrea Y. Connolly

Proofreader
Karen K. Johnson

Production Artist
Lezlly Freier

Indexer
Frances Bowles

Color Separations by
Creative Color

Lithographed in USA by
Webcrafters, Inc.

Consultants

Carlton Koehler, PhD.
   Department of Entomology
   University of California at
   Berkeley
Rex Marsh
   Department of Wildlife and
   Fisheries Biology
   University of California at
   Davis
William H. Robinson, PhD.
   Department of Entomology
   Virginia Polytechnic
   Institute and State
   University
   Blacksburg, Va.
Patricia A. Zungoli, PhD.
   Department of Entomology
   Clemson University
   Clemson, S.C.

Special Thanks

Lee Barone; Chula Camp; Warren Camp; Camp Brothers Home Remodelling, San Francisco, Calif.; Jacqueline Carrigan; Chula Productions, San Francisco, Calif.; Dominique Fleming; Wade Fujino; Marry Gilley; Jean Kasha; Strybing Arboretum, San Francisco, Calif.; Sudbury Products Division of Farnam Companies, Inc., Phoenix, Ariz.; Symons Nursery, San Jose, Calif.; Jean Zaija.

Photographers

Max Badgley: Front cover TL, 36B, 37B, 38, 47R, 48L, 74C, 74R, 88R, 89TL, back cover BL
Jerry Clark: 67T
James F. Dill: 42R, 56R, 79C, 84C, 90L
Walter Ebeling: 33B, 89R
Dr. Richard Elzinga: 88L
Charles Marden Fitch: 39R, 48R, 58, 66
C. Fryer/Australasian Nature Transparencies: 59BR
Herman H. Giethoorn/ VALAN Photos: 91T
Ted Granovsky: 28, 29B, 54, 60
Pam Hickman/VALAN Photos: 75
Saxon Holt: Title page, 4, 7T, 7C, 14B, 15T, 15C, 15B, 16B, 17TL, 17TR, 17B, 18TL, 18TR, 18B, 19L, 19R, 20, 22, 25, 27TL, 27TR, 27C, 30, 34L, back cover TL, BR
Frank E. Johnson/VALAN Photos: 71C
Ray R. Kriner: 35L, 44R, 53T, 55, 90R
Dwight R. Kuhn: 14T, 61, 83L, back cover TR
Martin Kuhnigk/VALAN Photos: 62
Wayne Lankinen/VALAN Photos: 71R
George D. Lepp/ Comstock: 81L
Stephen Marley: 6, 12
Richard W. Merritt: 71L, 87
Wayne Moore: 32, 34C, 34R, 46L, 53B, 59TR, 79R
Nelson-Bohart Assoc.: 42L
Pamela Peirce: 81R
J.R. Page/VALAN Photos: 65
Cecil Quirino: Front cover TR
William A. Robinson: 27B, 39L, 44L
Otto Rogge/Australasian Nature Transparencies: Front cover BR, 35C
Dennis W. Schmidt/VALAN Photos: 64
Runk/Schoenberger/Grant Heilman: 7B
David Scharf: 84R
Robert C. Simpson/VALAN Photos: 79L
Gerardus B. Staal/Zoecon: 13
Norm Thomas: 74L
USDA Forest Service: 16T, 29T, 52L, 52C, 52R
Van Waters and Rogers: 57R, 78L, 84L
Dave Watts/Australasian Nature Transparencies: 67B
Ron West: Front cover TC, C, BL, 33T, 35R, 36T, 37T, 40L, 40R, 41, 43L, 43R, 45L, 45C, 45R, 46R, 49, 50, 56L, 57L, 59TL, 59BL, 72, 76, 77L, 77R, 78R, 80, 81R, 82TL, 82BL, 82R, 83R, 84C, 86, 89BL, 91B, 92L, 92C, 92R
Val and Alan Wilkinson/ VALAN Photos: Front cover CL, 47L

Front cover
Top left: Bean weevils in mixed beans.
Top center: Black widow spider.
Top right: Ants feeding at bait.
Center left: Firebrat.
Center: Carpenter bee larvae.
Lower left: American cockroach with egg case.
Lower right: Mud dauber at nest.

Title page
Set snap traps in places where people won't accidentally step on them. If pets or small children are present, keep traps hidden from their view.

Back cover
Top left: Mouse bait placed in kitchen cabinet.
Top right: A common household spider.
Lower left: Bean weevil.
Lower right: Fly sticky trap.

CONTROLLING HOUSEHOLD PESTS

SAFE AND EFFECTIVE HOME PEST CONTROL

Learn how to keep pests out of your house, detect their damage, and use the most effective methods to eliminate infestations.

Page 5

KITCHEN AND HOUSEHOLD PESTS

Prevent ants, cockroaches, flies, and many other pests from sharing your home, stored food, and household effects.

Page 31

STRUCTURE AND WOOD PESTS

Timely prevention and control of termites and other destroyers of wood can save thousands of dollars in repairs.

Page 51

RODENTS AND OTHER ANIMAL PESTS

Wild mammals and some birds are nuisances if they nest in or on your home or get into stored food and garbage. Here's how you can discourage and trap them.

Page 63

BITING AND STINGING PESTS

Avoid stings and bites by controlling pests in both yard and home, and learn the correct first aid to use in case of injury.

Page 73

SAFE AND EFFECTIVE HOME PEST CONTROL

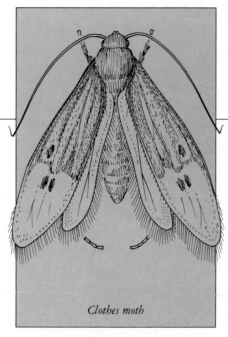

Clothes moth

Learn how to prevent household pests by denying them entrance, food, and shelter. If pest outbreaks do occur, here's all you need to know to use pesticides, traps, and baits safely and effectively.

Household pests have plagued people—destroying food, ruining personal property, and spreading disease—since the beginning of history. Millions of years ago, cockroaches probably made themselves at home in cave-dwellers' quarters just as comfortably as they do today in twentieth-century high-rise apartment buildings. Fabric pests undoubtedly ate holes in the animal-skin clothes of the Neanderthals just as they ruin our sweaters, carpets, and other woolens. During the Middle Ages, primarily in the fourteenth century, a third of Europe's population died from plague, spread mostly by fleas harbored by rats. Even if the cause had been known, no one would have been able to stop the disease, which now can be controlled with antibiotics.

As society evolved in the Western Hemisphere, household pests weren't far behind—we've even acquired some new ones. As recently as the beginning of this century, bed bugs, disease-carrying mosquitoes, and other biting pests were common in and around even the best-kept homes. Because it is now understood that unsanitary conditions and poor housekeeping can encourage pests, most people feel ashamed and somehow responsible when pests invade their homes. But they shouldn't blame themselves for most pest problems—wherever people live, pests will follow.

In cities, rats and cockroaches make no distinction between social classes; garbage is garbage to them, whether it is from the rich or the poor. Pests often flourish where poverty breeds, however, because

Quick and thorough action can control pest outbreaks in your home. Prediluted pesticides make the job easier.

of crowded, inadequate living conditions and lack of resources for control measures.

Until just a century ago, there were few effective ways of dealing with pests. Today that is no longer true. Although scientists still have much to learn about pests, they presently possess a staggering amount of knowledge about insects and other household pests, and new discoveries are being made every year.

As scientists have learned more about pests, they have discovered new ways to control them. During the last 50 years, new pesticides have been developed that are more effective than was ever before thought possible. For the first time in history, people needn't resign themselves to sharing their homes with household pests. A combination of good housekeeping and the proper use of the right pesticides can effectively and safely control destructive household pests.

WHY CONTROL PESTS?

To some extent a pest is a pest only in the eye of the beholder. People's reactions to different household pests vary widely—while one person may shriek at the sight of a spider, another may catch it and release it safely outdoors. For the most part, people simply don't want uninvited guests dwelling with them, even if they cause no damage. Nondestructive pests are called nuisance pests because they are annoying or embarrassing, although they are not harmful in the numbers usually found in homes.

Household pest problems become more serious when you discover that the pest is destroying your property. A termite infestation can easily cause thousands of dollars worth of damage before you discover it. Kitchen pests can ruin stored food, putting a dent in the household budget and causing a feeling of revulsion when wormy flour or beans are discovered. Carpet beetles and clothes moths show no regard for the value of a fabric and will destroy an expensive fur coat as readily as a forgotten woolen rag.

Household pests also transmit diseases. Rodents, cockroaches, and houseflies can cause food poisoning by

Many pests are attracted to moisture in bathrooms. Promptly fix any leaking pipes and dripping faucets.

the pest's venom. However, researchers believe that some of the deaths attributed to car accidents, sun stroke, or heart attacks may actually have been caused by insect stings.

This book discusses the pests that you are likely to find in and around your home. Each pest is discussed in detail in the next four chapters. The pests are grouped according to the type of damage they do: The second chapter covers pantry and household pests, the third chapter deals with pests that destroy wood, the fourth chapter discusses rodents and other animal pests, and the last chapter covers biting and stinging pests that can be found indoors or out. You will learn how to identify them and the damage they do as well as the most effective ways of controlling these annoying and damaging insects and animals.

The remainder of this first chapter deals with general pest control methods. Don't overlook the important information it contains. You will read about how to exclude pests from your home, how to detect their damage, and how to handle pesticides and traps safely and effectively.

AVOIDING HOUSEHOLD PESTS

You can prevent many pest problems from ever occurring by following the suggestions given here. In fact, it is sometimes easier to prevent pests from infesting your home than it is to get rid of them once they become established. For instance, termite infestations can go unnoticed for years. The damage is often discovered only when you sell your home, and by then the infestation may have done enough damage to reduce the market value of your house considerably. An annual professional termite inspection, which catches any termite damage early, usually costs less than fifty dollars. Early treatment costs only a few hundred dollars, not much compared to the thousands needed to fix extensive damage.

You can easily carry out most pest-prevention steps yourself or have them done by a repair person. Some pest-prevention measures, such as household repairs or improvements, need be done only once; then you can forget about

spreading salmonella bacteria. Roaches have also been suspected of carrying toxoplasma parasites, which cause a mild, flulike illness in most adults, but which can cause birth defects in an unborn child if a pregnant woman comes down with the disease. By carrying disease organisms, pests have spread many of humankind's worst killer diseases— malaria, typhus, and bubonic plague— which have killed many millions of people over the centuries.

Many insect- and rodent-borne diseases still abound in developing countries, but pests can cause illnesses even in North America today. Allergies to cockroach particles and feces in food and household dust may be a primary cause

of asthma in people. Ticks can transmit Lyme disease and Rocky Mountain spotted fever—serious, sometimes deadly illnesses. Encephalitis is the only mosquito-borne disease currently of serious concern in the United States; this viral disease affects thousands of people each year. Mosquito-borne malaria is prevalent in areas of Asia, Africa, and South America and still crops up in the Caribbean; it could establish itself again in the United States if brought in by foreign travelers.

Venomous stinging and biting pests, including scorpions, bees, spiders, and wasps, inflict millions of bites and stings in the United States each year. Fewer than 100 people per year die as a direct result of such stings, almost always because the victim was unusually allergic to them, knowing that you have made your home more pestproof. Other steps, such as keeping crumbs swept up, require diligence every day. In either case, you can expect that the more preventive steps you follow, the fewer pest problems you are likely to have. If pests enter your home even after you have taken preventive steps, you can generally control the problem more easily with insecticides or traps than if you hadn't established a sound pest-avoidance program in the first place.

Preventive techniques are not purely preventive; they help control pests, too. Make the following procedures an integral part of your pest control efforts, together with pesticides or traps designed to control particular pests. A good pest-avoidance program is based on a three-pronged plan of attack:
• Take away points of entry.
• Take away food and water.
• Take away shelter.
Of course, the entry points, food, and shelter of particular pests may differ.

No Admittance

Keeping pests out of your house is the cornerstone of a pest-prevention strategy. Many pests, such as millipedes, centipedes, cockroaches, spiders, earwigs, and ants, enter your home through cracks in the foundation or siding; others, such as mosquitoes and flies, get in through broken screens; and still others, such as mice, rats, and bats, may gain access through broken siding, unscreened vents, and chimneys. Your job is to seal all of these entry spots.

Above: One way to prevent pests from entering the house is to seal off their entry spots. Here, foam-rubber weather stripping installed under the metal threshold and caulking applied to cracks at the sides of the doorsill help keep out crawling insects.
Center: Discourage pests from taking up residence by leaving them nothing to feed upon. For instance, if food doesn't come in pest-safe containers, repackage it before storing.
Below: One way to discourage wood-boring pests is to move their nesting sites. Here, firewood is stacked more than 10 feet from the house so that wood-boring pests will be unable to move easily from the firewood to the structural wood.

Using a Caulking Gun

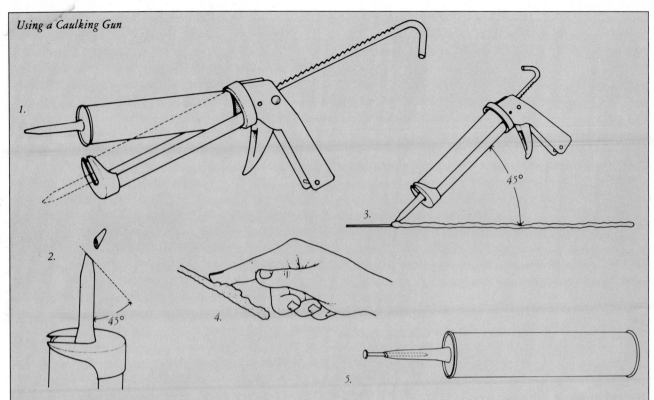

1. Pull the plunger all the way back and insert the cartridge, seating the bottom end first. Squeeze the trigger a few times until the plunger makes contact with the tube.

2. Cut off the tip of the nozzle at an angle; the closer to the tip the finer the bead. Start with a fine bead and enlarge it if necessary. Puncture the seal inside the nozzle with a long nail or stiff wire.

3. Force caulk into the crack by pushing the gun forward at an angle and squeezing the trigger. The caulk should be as deep as the width of the crack and should adhere firmly to both sides. Stop pulling the trigger and release the plunger just before you reach the end of the crack.

4. Moisten your finger with soapy water (for latex caulk) or mineral oil (for other types of caulk) and smooth the caulk and clean off adjacent surfaces.

5. Release the plunger and insert a nail into the nozzle. Wipe off any excess caulk and cover the nozzle with foil or plastic.

Where to Caulk a Home

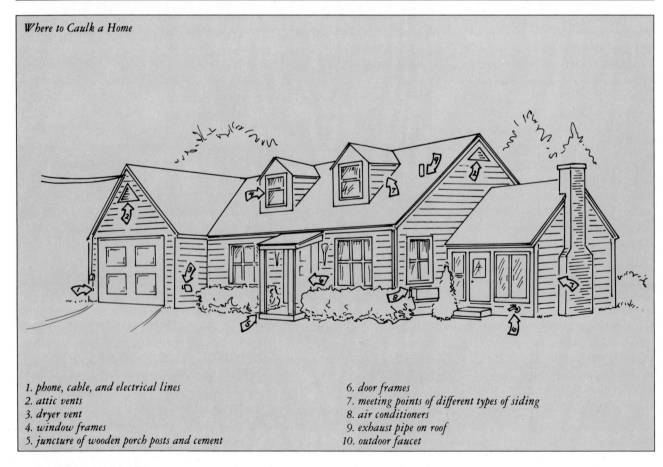

1. phone, cable, and electrical lines
2. attic vents
3. dryer vent
4. window frames
5. juncture of wooden porch posts and cement

6. door frames
7. meeting points of different types of siding
8. air conditioners
9. exhaust pipe on roof
10. outdoor faucet

Examine the outside foundation of the house and the siding for cracks—even hairline ones—that have occurred as the house ages and settles. Look for holes in the siding where electrical, television, or telephone lines or plumbing pipes come inside. Finally, see if there are any cracks around the frames of doors and windows. Seal all of these cracks and gaps with caulk or weather stripping. Not only will caulking help keep out pests, it is an important home maintenance procedure that prevents moisture buildup and subsequent wood decay.

You can choose from many different types of caulking materials; the one to use depends on the size of the crack. Be sure to select a high-quality product, and apply it with a caulking gun. The best time to caulk is during warm weather, when the caulking material is more pliable and easier to work with. (See *Ortho's Home Improvement Encyclopedia* for more information on caulking.)

Cracks around doors are one of the most common ways that insect pests get into homes. Properly installed weather stripping on the door and door frame is all you need to seal off these entry points.

Attic vents and chimneys are other possible entry points, especially for animal pests such as raccoons, bats, roof rats, and birds. Be sure that any vent holes and chimneys on the roof or sides of the house are screened to prevent pests from entering. Do not use fine-mesh screen, such as the type used for window screens, on chimney vents, because it will slow down the free flow of air. The best material to use is 1-inch-mesh hardware cloth. A spark-arrester screen installed on top of the chimney will both keep animals and birds out and help prevent accidental fires. Also, be sure that the exhaust pipe from the clothes dryer has a flap that works properly and is not clogged with lint.

Window and door screens provide a simple and effective way of reducing the number of flies, bees, and mosquitoes that come indoors. Be sure that your house has tight-fitting screens that are in good repair. A screen that has space around the edges allows small insects to get in. If there are tears in any screens, you can repair the tear or replace the entire screen (see below and page 10).

Sheet metal and concrete are materials that rodents can't gnaw through, so they are good choices for sealing possible entry points around pipes.

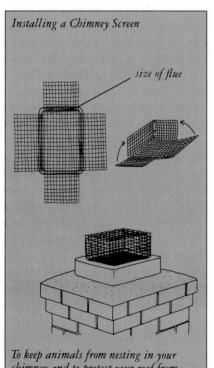

Installing a Chimney Screen

size of flue

To keep animals from nesting in your chimney and to protect your roof from sparks, install a chimney screen. Use heavy galvanized mesh with a 1-inch grid. Cut and fold the mesh into a rectangular shape that will wedge into the opening of the flue.

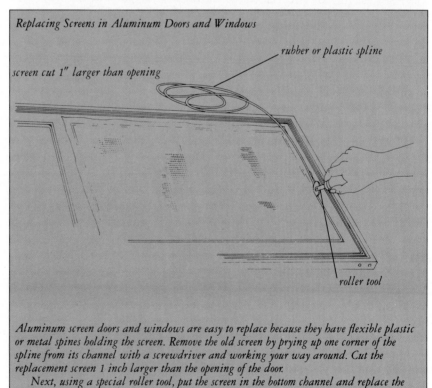

Replacing Screens in Aluminum Doors and Windows

rubber or plastic spline

screen cut 1" larger than opening

roller tool

Aluminum screen doors and windows are easy to replace because they have flexible plastic or metal spines holding the screen. Remove the old screen by prying up one corner of the spline from its channel with a screwdriver and working your way around. Cut the replacement screen 1 inch larger than the opening of the door.

Next, using a special roller tool, put the screen in the bottom channel and replace the spline. Keeping the aluminum door perfectly square, work your way all the way around the screen. The pressure from the tool and from the spline compressing into the channel will pull the screen taut.

Finally, trim off excess screen using wire cutters.

Place planks over a pair of sawhorses to provide a platform where you can clamp the frame of the screen.

1. Pry up the molding (screen bead) around the screen, being very careful not to break it.

2. Use a screwdriver or the corner of a putty knife to remove the staples holding the old screen.

3. Cut a piece of new screen so it overlaps the opening by 1 inch all around. Attach the screen across the bottom edge of the door only, using a staple gun. Be sure the screen is straight and the staples are snug.

4. Stretch the screen by bending the door frame slightly with boards placed under each end and clamping the middle to the sawhorse planks.

5. Staple the top of the screen in place, release the tension slowly, then staple both sides. Staple the center rail last.

6. Trim the excess screen with a sharp knife and replace the molding.

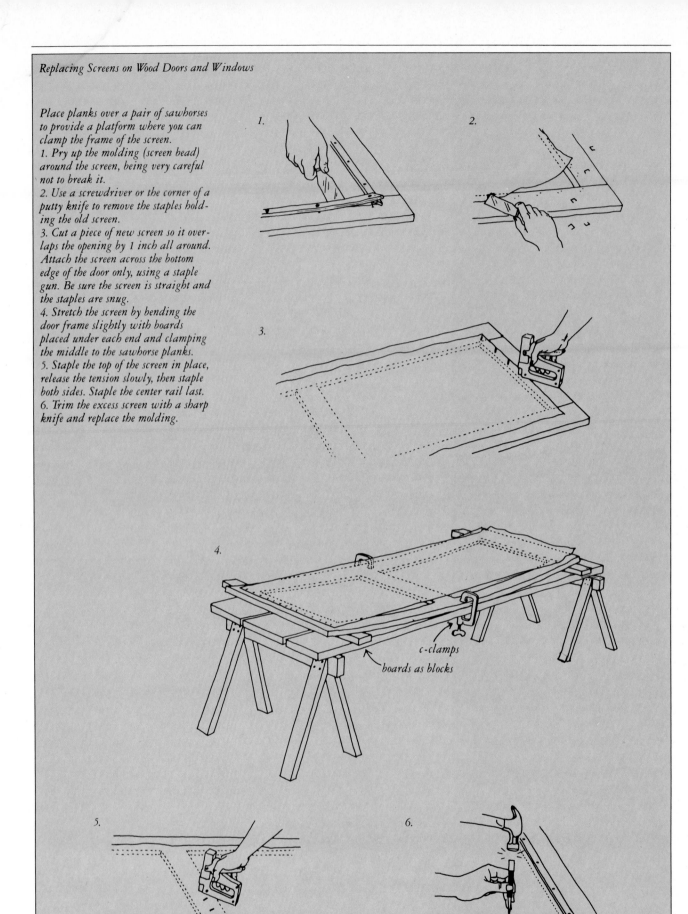

c-clamps

boards as blocks

Pests sometimes get into homes by hitchhiking on new and used appliances and furniture, grocery bags, or cardboard boxes. Inspect these items before you bring them indoors.

No Vacancy

Even if pests do get into your house, they may become discouraged and seek more hospitable accommodations if there's no place for them to nest or hide. Your strategy is either to remove pest hiding places—both indoors and out—or to make those hiding places uncomfortable. Most crawling pests need dark, tight places in which to hide during the day. Caulking cracks along baseboards, walls, cabinets, and elsewhere will eliminate many potential hiding places.

Stacks of magazines, newspapers, and cardboard boxes in basements and garages make good hiding places, and they can also provide pests with food. Even if newspapers are stacked neatly, mice may nest in them, silverfish may eat them, and roaches may hide between the pages. Don't allow these items to accumulate, especially if you have pest problems.

The areas under and around garbage cans are prime breeding places for certain pests. Removing nearby boards and other debris from the ground will help keep them away. Treating the ground around the garbage cans with a pesticide is also a good way to prevent a buildup of pests.

Some pests, including millipedes, psocids, sowbugs, and many wood-destroying insects and fungal rots, require moist conditions to survive. All you may need to do to discourage them is eliminate any damp areas and repair any leaky pipes. These and other pests, such as spiders, can breed in accumulated plant litter, mulch, or firewood stacked around the house. Keep the area next to the foundation free of accumulated leaves, compost piles, and similar items. If you plant shrubs next to your home, keep them trimmed back so that they do not touch the house. Do not mulch heavily around foundation plants, because mulch provides an ideal breeding place for many household pests. Always store firewood away from your home. (See page 55 for details on firewood storage.)

You can spray or dust insecticide on any potential hiding places that you can't eliminate to make them inhospitable to pests. If the area is dry, some pesticides, such as boric acid powder and silica aerogel, will remain active for many months. Residual sprays that last several weeks or more include Dursban® (chlorpyrifos), diazinon, and Baygon® (propoxur).

No Food or Drink Allowed

General cleanliness is the key to starving pests out of your house, but to eliminate pests completely, you'll need to go beyond plain good housekeeping. Even immaculately clean homes may be infested with cockroaches, ants, and other kitchen pests. Open boxes of cereal, crackers, and pet food lined up neatly in a clean pantry provide pests with just as much food as a messy cupboard does. Cakes, cookies, and breads sealed only in plastic wrap and left on the counter or in a drawer can be a banquet for insects and rodents, which will chew right through the wrapper. Night-active insects feeding on these items may easily go unnoticed.

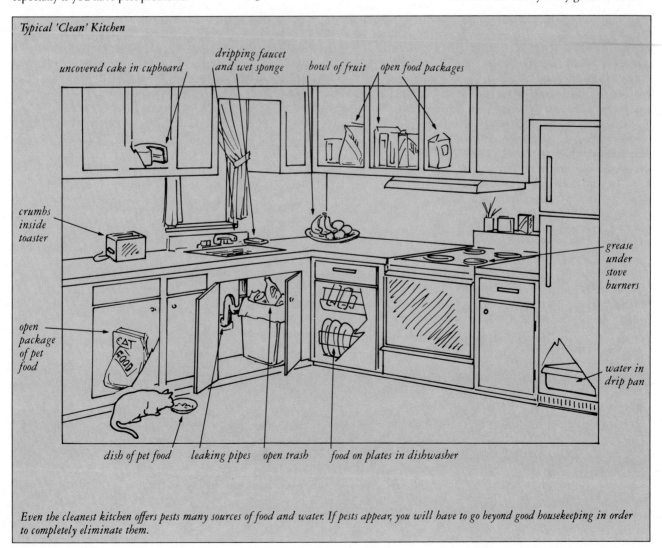

Typical 'Clean' Kitchen

uncovered cake in cupboard

dripping faucet and wet sponge

bowl of fruit

open food packages

crumbs inside toaster

grease under stove burners

open package of pet food

water in drip pan

dish of pet food   leaking pipes   open trash   food on plates in dishwasher

Even the cleanest kitchen offers pests many sources of food and water. If pests appear, you will have to go beyond good housekeeping in order to completely eliminate them.

Pests will, of course, be more numerous wherever housekeeping is lax. Crumbs left on the kitchen or dining room floor, kitchen garbage cans with open lids, and spilled food in cupboards, drawers, and stoves offer ample, inviting feasts for foraging pests.

If pests are a problem—or if you want to prevent them from becoming one—seal food stored in the cupboards or pantry so that they can't get to it. Storing flour, cereal, and similar foods in plastic, glass, or metal containers with tight-fitting lids prevents cockroaches, flour moths and beetles, ants, and similar pests from infesting the foods. Plastic wrappers and cardboard boxes will not keep them out. Store pet food in a similar manner to keep out rodents and insects. One way to store a large sack of dry dog food is in a plastic garbage can with a snug lid.

Don't keep flour and dried foods for long periods of time, even when properly stored—purchase only as much as you will use within a month or so. By constantly rotating stored foods, you eliminate or prevent pest problems caused when boxes or bags of food are infested when you bring them home from the store. An infested package may contain only eggs or a few larvae and thus be difficult to detect; this level of infestation poses no health problems.

Water is just as important to certain pests as food. If these pests can't find water, they may die. Common sources of water include liquid left in empty soda and beer cans, leaking plumbing or dripping faucets, and pet water bowls. Pests will also drink from spills on kitchen counters and water in dish drainers, as well as from soggy sponges left near the sink. Water can also accumulate in the drip pan of a frost-free refrigerator if it doesn't evaporate fast enough; check this pan regularly, and empty it if necessary.

FIGHTING PESTS WITH PESTICIDES

A pesticide is any chemical agent used to control pests. For instance, table salt acts as a pesticide when you sprinkle it on snails and slugs. The most common use of pesticides is in controlling a pest outbreak that is already underway, but in some cases pesticides can prevent pests from building up to damaging levels in the first place. Spraying pesticide around any openings in the house that you can't seal, and spraying or applying a granular insecticide around the outside of your home's foundation, will greatly reduce the number of pests coming indoors.

The different kinds of pesticides are named according to the type of pest they control, as follows:
- Insecticides control insects.
- Fungicides control mold and fungi.
- Herbicides control weeds.
- Miticides control mites.
- Rodenticides control rodents (mice and rats, for example).
- Molluscicides control mollusks (snails and slugs).
- Bactericides (these include household cleansers) control bacteria.

Insecticides

Because most household pests are insects or spiders, insecticides are the most common pesticide you will need around your home. Insecticides act primarily as either contact poisons or stomach poisons. Contact poisons work when they are sprayed directly on an insect or when the insect walks across the residue. Stomach poisons must be ingested to be effective. The insect often ingests the poison when it grooms itself and licks the pesticide off its appendages or when it feeds on pesticide-treated bait or food.

A clean kitchen can stymie most pests, although pests may still be carried in on purchased food, or wander in from outdoors.

The majority of the insecticides available kill insects by interfering with the functioning of their nervous systems. Even the natural insecticide pyrethrins does this. One advantage of this type of insecticide is that it kills the insect very quickly, which is what most people want. Another is that it is effective in small quantities. These insecticides work by overloading the nervous system. The insect first loses the ability to coordinate its muscles and glands. You may have seen an insect's legs twitching after it was sprayed. This happens because the nervous system is sending continual messages to all of the insect's muscles. The muscles controlling respiration quickly cease to work, and the insect soon dies. The whole process may take less than a minute if the insect receives enough of the right insecticide.

Pests can also be killed by the clogging of their breathing pores. Insects and related pests, such as mites and spiders, do not have lungs. Rather, they breathe through a network of tubes called tracheae, which carry air to different areas of the body. Air enters the tracheae through numerous small openings in the body surface. The most common suffocating substances are spray oils. When you apply one of these oils, it clogs the tracheae, causing the pest to suffocate quickly.

Desiccants cause insects to dry out faster than they can take in water. They work by damaging the waxy or oily film that coats an insect's body and prevents it from drying out. Boric acid powder and silica aerogel dust are the two most common desiccants used to control household pests.

Insecticides can be classified according to the chemical group to which they belong. The first synthetic insecticides, chlorinated hydrocarbons (or organochlorines), were developed in 1939 when the Swiss entomologist Dr. Paul Müller discovered the insecticidal properties of DDT. He later received the Nobel prize in medicine for this discovery. DDT was first used during World War II and quickly began to play a key role in the control of mosquitoes, lice, fleas, and other insects that transmit human diseases. It saved millions of lives and helped greatly reduce the amount of food lost to pests. However, scientists suspected this compound of accumulating in the food chain and being detrimental to certain forms of wildlife. It was banned in the United States in 1973 but is still used today in many developing countries, where its benefits outweigh any environmental risks. In the 1950s and 1960s, a number of other DDT relatives were discovered, most of which are unavailable today.

Organophosphate insecticides were developed next. Unlike some of the chlorinated hydrocarbons, organophosphates are short-lived insecticides and do not build up in the food chain. Many of the insecticides in use today, including Orthene® (acephate), Dursban® (chlorpyrifos), diazinon, and malathion belong to this group.

Carbamates were first introduced in 1951. They are very similar to organophosphates but are derived from carbamic acid rather than phosphoric acid. The two most common carbamates are Sevin® (carbaryl) and Baygon® (propoxur).

Botanical insecticides are poisons extracted from plants. Pyrethrins, nicotine, and rotenone are common examples. These are not necessarily safer to use than the other insecticides. In fact, nicotine is fairly toxic. Pyrethrins is the botanical insecticide used most often. This nerve poison is extracted from the flower heads of a chrysanthemum relative grown primarily in South America and Africa. Insecticides related to pyrethrins have been synthesized in the laboratory and are now used in a wide variety of products. These last longer than the natural substance and are more effective.

A few other classes of insecticides contain chemicals useful for controlling household pests. Inorganics are compounds that do not contain any carbon. They include boric acid, silica aerogel, and sulfur, which act partly as desiccants. Fumigants, such as methyl bromide and sulfuryl fluoride, become gases at temperatures above 40° F; the pests breathe the pesticide and are killed.

Largely because of our experience with DDT, pesticides today are more closely regulated than almost any other chemical. The Environmental Protection Agency (EPA) registers all pesticides and determines that they are safe and effective for their labeled uses.

Growth Regulators

When used as pesticides, insect growth regulators (IGRs) disrupt the insect's growth process by mimicking hormones that regulate molting and metamorphosis. Because these growth processes are unique to insects and their relatives, IGRs are very specific in their toxicity. They are some of the safest pesticides you can apply.

The most common IGR is methoprene (Altosid®). This insecticide mimics a key hormone that regulates when an insect will molt its skin and when it will metamorphose into a pupa. Used as a pesticide, it causes insect larvae to grow larger and molt repeatedly. They never pupate into reproductive adults. Eventually, the larvae die.

Methoprene can be used to control mosquitoes and fleas, both of which are pests only in the adult stage. Neither bite in the larval stage, and mosquito larvae are even useful, since they are an important source of food for fish.

Other IGRs work in slightly different ways. Hydroprene prevents the reproductive systems of cockroaches from developing, resulting in sterility. Dimilin®, used in agriculture, interferes with the hardening of the insect's skeleton and mouthparts, causing the insect to die of starvation.

Because growth regulators do not provide immediate relief from insects, IGR sprays may also contain conventional insecticides to kill off the current pest population. The IGR lasts longer than the other insecticide and continues to control new insects that hatch or migrate into the area.

Also available is a growth regulator for rodents. It works by causing sterility in male rats and mice.

At left is a normal adult American cockroach and at right a cockroach made permanently immature by treatment with a growth regulator.

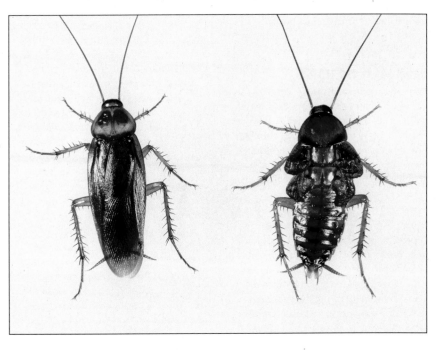

Microbials

Microbials kill by causing a fatal disease in insects. These insecticides are not chemicals but rather microorganisms, such as bacteria or viruses. The two most common microbials in use today control garden pests. They are *Bacillus thuringiensis*, which kills only larvae (caterpillars) of butterflies and moths, and *B. popillae*, which kills the grubs of Japanese beetles. *B. thuringiensis* var. *israelensis* is a newly developed variety that affects mosquito larvae.

These disease organisms are very specific, affecting only certain groups of insects. Microbials are very safe to apply because they can't harm people, pets, or nontargeted organisms. But because they are so specific, if you are trying to control several types of pests, you will need to apply more than one pesticide. Microbials also do not act as quickly as conventional insecticides and therefore are not useful in situations where you desire a quick kill.

Rodenticides

Rodents are probably second only to insect pests when it comes to home pest problems. Many rodenticides, such as warfarin, pindone, diphacinone, chlorophacinone, and brodifacoum, are anticoagulants, killing rodents by causing them to bleed internally. These are usually prepared as baits, which the rodents eat over a period of several days. Other rodenticides act as poisons and kill the rodents more quickly than anticoagulants. These include bromethalin, cholecalciferol, red squill, strychnine, and zinc phosphide. All rodenticides can harm people, pets, or other animals if ingested and thus must be used in strict accordance with label directions.

PESTICIDE FORMULATIONS

Pesticides come in a variety of forms. These different formulations help make them easier to apply, more effective, or less hazardous to handle. The formulation you choose depends on the equipment you have, your particular pest problem, and your personal preference.

Sprays

Most pesticides are diluted in water or another solvent and sprayed on areas to be treated. They are frequently available as a concentrated liquid containing the active ingredient, water or another solvent, and usually an emulsifying agent that helps it mix easily with water. You measure the concentrate and combine it

with water before spraying it. After being sprayed, the solvents and water evaporate, leaving the insecticide residue, which adheres to the sprayed surface.

Liquid concentrates are the most common formulation for outdoor use. For indoors you can buy prediluted pesticide sprays in handheld plastic pump dispensers. These are convenient to use and apply because you don't need spray equipment and you don't have to measure and mix the pesticide.

Aerosols

Aerosols are a mixture of the active ingredient, a solvent, and a gas under pressure. Available in ready-to-use spray cans, aerosols require no mixing or special equipment, and they don't lose their strength while in the container during their normal period of use. They also

Above: Rodenticides are one weapon against invasions of rodents like this deer mouse. Below: This economy-sized container of prediluted insecticide comes with an easily attached tube and trigger sprayer. It is ideal for jobs such as spraying baseboards and other insect hiding places.

produce an extremely fine spray mist that makes them useful for controlling flying insects. These characteristics have made aerosols one of the most popular formulations for indoor use.

Total-release aerosols—sometimes called bombs—give off a very fine, far-reaching spray that can blanket an entire room with pesticide. These are commonly used to control pests, such as fleas, that hide in cracks and crevices.

Dusts

Dusts are finely ground, dry mixtures combining a low concentration of pesticide with an inert carrier, such as talc, clay, or volcanic ash. They are easy to use, requiring no mixing. Although they are not easily absorbed by the skin, dusts can be inhaled if blown about, so apply them only when the air is still.

Dusts are useful for treating wall voids and the galleries of carpenter ants and other wood borers, because the dust drifts around in the space, providing more thorough coverage than is possible with a spray. Dusts do not stain surfaces, as some sprays can, but they do leave a visible residue on any surface you treat. In many situations, the messy look of this residue is undesirable.

Granules

Like dusts, granulated formulations contain the active ingredient mixed with an inert carrier. Granules solve one of the main disadvantages of dusts—the problem of being easily blown away. They retain the advantage of not being easily absorbed through the skin. Granules are used primarily as a barrier around foundations to control crawling household pests.

Wettable Powders

Wettable powders are similar to dusts. They are made up of an active ingredient plus an inert carrier, but in addition they contain a wetting agent, which allows them to mix readily with water, and a spreader-sticker, which helps them spread on the sprayed surface and stick to it. Wettable powders settle in the spray tank fairly quickly and must be agitated constantly. They also may wear out spray equipment faster than liquid concentrates do because the carriers tend to be abrasive, and they are also more likely to clog screens and nozzles. Because of these

Above: A ready-to-use container of insecticide dust makes it easy to penetrate small cracks. Always clean up exposed dust, so children and pets won't touch it. Center: Applying insecticide granules around a house foundation helps control many types of pests that can enter homes. A 3- to 5-foot band of insecticide is usually enough. Below: Place rodent or roach bait where pests will find it, but where it is not accessible to children and pets.

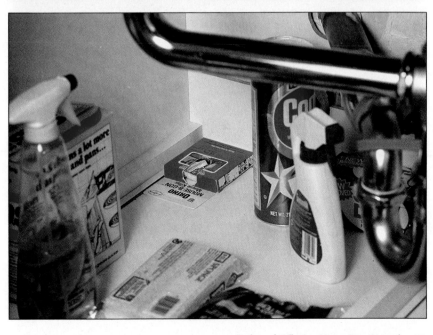

Tent fumigation ensures that an entire drywood termite colony will be exterminated.

drawbacks, only a few pesticides—ones that work best that way—are formulated as wettable powders.

Baits

Baits, commonly used to control cockroaches, ants, and rodents, contain a pesticide mixed with a food that the pest likes. Sometimes additional attractants are added to the mixture to increase the chances that the pest will find the bait and eat enough of it for good control. Most baits used for household pest control do not attract pests from far away. With cockroaches, for instance, you must rely on their random crawling behavior to get them within a few inches of the bait before they can detect it. For this reason, baits are most effective when placed where the pests tend to gather or search for food. They also work better when other suitable food is lacking.

Baits are convenient because they do not require special application equipment and they usually come ready to use. Since you are relying on the bait to draw the pest to the pesticide, rather than relying on the pest to walk across a pesticide residue by chance, baits involve less pesticide than sprays or dusts. Baits must be used carefully so that they are not accessible to children or pets.

Fumigants

Fumigants are pesticides that are applied as gases, but they can be stored in solid, liquid, or gas form. Because they are gases, fumigants are able to penetrate into much smaller crevices than fogs or

mists. Their gaseous form also makes them easier to inhale and thus more dangerous to work with than other types of pesticides. Only licensed pest control operators are allowed to use fumigants in buildings. The most common use of fumigants indoors is to control drywood termites and wood-destroying beetles.

The gas is contained within a tent or other barrier. After the tent is removed and the structure ventilated, the fumigant dissipates, leaving no hazardous residue. However, it also does not leave any protective effect, and so reinfestation is possible soon after treatment.

Encapsulated Pesticides

Flea collars, resinous pest strips, and other time-release or encapsulated pesticides contain an active ingredient in a porous plastic membrane that allows it to escape slowly. Because the pesticide is released slowly, the product controls insects over a long period. Encapsulation is an expensive process, which tends to limit its use.

Encapsulated pesticides are formulated for use in a variety of different situations. Pest control strips are very effective against flying insects in confined areas. The insecticide in these strips is usually dichlorvos, which escapes slowly as a gas. Dichlorvos is also commonly used in flea collars, which work like the impregnated strips.

Encapsulated diazinon can be applied as a spray to baseboards and other areas where you want a long residual effect. When formulated as a bait, it is useful for controlling yellowjackets and other social insects. The slow-release formula allows a foraging yellowjacket to share the bait with other members of the colony before it is killed.

APPLICATION EQUIPMENT

In most cases, you will control household pests with pesticides sold in ready-to-use containers. Handheld pump dispensers and aerosol spray cans are convenient because they require no mixing or diluting and because they are ready at a moment's notice. In some situations, however, you may want to apply a pesticide to a large area of your home; in such instances, it may be more economical to purchase a concentrated pesticide and apply it with special application equipment. The following applicators are most useful for controlling home pests.

Sprayers

A hose-end sprayer is a spray head above a glass or plastic jar that screws onto the end of a garden hose. Water pressure suctions up a concentrated pesticide/water mixture from the jar and mixes it to the proper concentration in the spray stream. Hose-end sprayers are not designed for indoor use because they deliver a large volume of spray very quickly. But you will find one useful for controlling such pests as fleas, ticks, cockroaches, earwigs, sowbugs, and boxelder bugs before they find their way indoors. You can also use a hose-end sprayer to spray around the house foundation to control ants, spiders, and similar pests.

Hose-end sprayers are convenient and inexpensive. Be sure to choose a quality sprayer that has an on-off valve and an adjustable nozzle.

A hose-end sprayer is useful for spraying the outdoor habitats of a number of outdoor pests that can become indoor nuisances.

Ready-To-Use Applicators

Many household pesticides come in a container that also serves as a dispenser. These are convenient because they don't require any additional application equipment; there's no need to measure, mix, or transfer the pesticide to another applicator, and there's no need to clean the applicator. You just throw it out when the contents are gone.

Ready-to-use pesticides are especially handy for indoor use because you often need only a small amount at a time. In such cases, it is impractical to mix up a gallon of spray, and the equipment used for outdoor applications is not suitable for this type of situation.

Aerosols

One of the most popular ways to apply an insecticide, especially indoors, is from an aerosol container. Aerosols contain insecticide mixed with a solvent and a gas under pressure. When you press the valve, the insecticide is dispersed in a fine mist and the solvent evaporates immediately. The label specifies whether a given aerosol is to be used to spray surfaces or spaces. Aerosols dispense such a fine mist that they very efficiently use a small amount of insecticide.

Total-Release Aerosols

These popular aerosols, sometimes called bombs, have a special nozzle that you set to automatically provide a continuous release until it has emptied its contents. Because you leave the room while it is spraying, you don't expose yourself to the mist. You must use the proper number of containers for the amount of space you need to treat; otherwise, you won't get effective control. Read the label to determine how many containers you need.

Above left: Pump dispensers are handy for treating small surface areas and are ready to use when you need them.
Above right: Aerosol containers dispense small amounts of insecticide very efficiently. Some aerosols can be sprayed in the air to control flying insects.
Below: Before setting off a total-release aerosol, move food and utensils to another room and close windows and doors.

Pump Dispensers

Many pesticides come in ready-to-use pump or trigger dispensers similar to those used for window-cleaning fluids. These are excellent for spot-treating surfaces, cracks, and crevices in and around your house. The nozzle often adjusts from a fine spray to a far-reaching jet. When you need to spray a space rather than a surface, use an aerosol, which produces a finer spray droplet and expels it with more force.

Dusters

Some dusts are sold in shaker cans made of cardboard with small holes in the top. You simply shake the cans to sprinkle the dust where it is needed. These cans are very convenient for small outdoor jobs. Dusts are also packaged in ready-to-use squeeze bottles. Both types are convenient, but neither expels the dust as evenly or as far as a plunger or crank duster (see photographs on page 19).

Granule Shaker Cans

Granules are also sold in cardboard shaker containers. Simply shake the can to dispense the granules. These are useful for covering small areas indoors and out (see center photograph on page 15).

Above left: A 1-gallon com-
pressed-air sprayer, filled with
the pesticide of your choice, can
be aimed with great accuracy.
Above right: You can adjust the
nozzle of a compressed-air
sprayer to produce a fine spray to
evenly cover a surface.
Below right: A coarse spray is
more far-reaching and can be
aimed into crevices.

A compressed-air sprayer consists of a tank with a plunger that you pump to create pressure in the tank. It usually holds 1 to 3 gallons of diluted pesticide. You can easily adjust the nozzle to produce anything from a fine mist to a coarse jet of spray, which allows you to use a compressed-air sprayer with great accuracy, an important feature if it will be used indoors. Some models have a wide or funnel top that makes for fewer spills when you pour pesticide and water into the tank.

The tank itself can be of plastic, galvanized steel, or stainless steel. Stainless steel tanks cost more than the others, but they last much longer and will not corrode. However, plastic tanks have important benefits: They are less expensive, lighter, and less likely to corrode than galvanized steel. They are also often somewhat translucent so that you can see the liquid inside. Galvanized steel sprayers tend to corrode quickly if you are not consistent in cleaning them thoroughly after each use.

An atomizer has a bicycle-style pump the tip of which is mounted on top of a small pint or quart tank that resembles a tin can or glass jar. The tank holds diluted spray. Single-action atomizers force air directly over a siphon tube, which draws the spray material from the container and atomizes it as it leaves the tube. This type of sprayer is most convenient for indoor use because it is good for small areas.

You can use a handheld pump sprayer or mister designed for misting houseplants as a pesticide applicator. These sprayers are usually inexpensive and made of plastic; they are handy for mixing up a small amount of pesticide to treat a few cracks and crevices indoors. Pump sprayers tend to wear out faster than well-made atomizers. Mark the bottle clearly if you use it for pesticides, and do not use it for other purposes.

Sometimes a paintbrush is the ideal tool for applying liquid pesticide indoors. Dip the brush in a container of diluted insecticide, and use it to treat baseboards, cracks and crevices, and similar areas around your home. It takes longer to apply an insecticide with a paintbrush than it does with a sprayer. Be sure to clean the paintbrush and container thoroughly after each use and clearly mark them both "for pesticide use only." Don't use them for any other purpose.

Dusters
Inexpensive plastic squeeze bottles designed to dispense catsup or mustard make handy dusters for household use. They are good for getting dusts directly into wall voids and cracks and crevices.

Above left: A commercial product modeled on the mustard-type squeeze bottle allows easy dispensing of insecticide dust indoors.
Above right: A handheld broadcast spreader works well when applying granules near the foundation, especially in ground cover where a wheeled spreader cannot be used.

Because these kinds of containers are normally used for food, you should clearly mark any you use for pesticides and store them safely with your other pesticides.

Pest control operators often use a bulb duster, which is similar to a squeeze bottle duster except that the part you squeeze is more bulbous and the tip has a metal attachment. These dusters can inject the dust with a little more force than catsup dispensers can.

A plunger duster has a bicycle-style pump that is mounted on a metal can. A crank duster has a small hopper or container for the dust and a hand-operated rotary crank. Both expel too much dust for indoor use.

Granule Applicators

A drop spreader looks like a small wheelbarrow. The hopper has a series of openings in the bottom through which pesticide granules, fertilizer, or seeds drop as you push the spreader. An agitator bar in the bottom of the bin, attached to the wheel axle, keeps the material flowing. You can adjust the size of the openings to control the amount of material flowing out. You may need to

use such a spreader outdoors for controlling lawn pests, such as ants, fleas, and ticks, that invade homes.

A broadcast spreader can be either handheld or wheeled. The pesticide granules, fertilizer, or seeds drop onto a rapidly turning wheel, or impeller, which flings the material several feet to the front and side as you walk or turn the crank. You can cover more area in a shorter period with a broadcast spreader than you can with a drop spreader, but you have less control over where the material lands.

You can modify the amount of pesticide you apply by adjusting the opening in the bottom of the hopper and by changing the speed at which you walk or turn the crank. Handheld spreaders are useful for applying smaller amounts of material and for applying material in ground covers and other areas where you can't push a wheeled spreader.

CARING FOR YOUR EQUIPMENT

Most application equipment will give you years of service with just a little time spent to maintain it. The most important step is to clean an applicator after each use, because some pesticides can be corrosive. Wash the equipment outdoors, not in the kitchen sink. Rinse the container at least three times with clean water, emptying the rinse water on the ground in an out-of-the-way corner of the yard. Spray some water through the nozzle during the last rinse to clean out the hose and nozzle. After the last rinse,

pour out the liquid, point the spray nozzle at the ground, and hold the tank high above it to drain any liquid still in the hose. Store the tank upside down so that it can drain.

Occasionally, rub a light coat of oil on any moving parts and gaskets. For compressed-air sprayers, oil the rubber portions of the plunger unit and squirt some oil inside the plunger to lubricate the main gasket.

Most sprayer problems start when foreign material clogs a nozzle. Begin taking care of your equipment by cleaning the sprayer—even a brand-new one—before each use. The manufacturing process may have left behind metallic chips and debris, which can clog a wand or nozzle and cause parts to wear, or bits of rust and dirt may have gotten in during storage. Flush the sprayer with fresh water; make sure screens and nozzles are clean.

While using the sprayer, check every so often to see whether any dirt, sand, or grass has clogged the nozzle. If the spray nozzle does become clogged, use a toothbrush or the cleaning pin that may have come with the sprayer to clean it. Anything else may widen the nozzle, ruining its spray pattern. If a nozzle seems plugged, back flush it with air or water (but don't put it in your mouth and blow through it!) or soak it in hot water. Remove the screen inside the nozzle occasionally and rinse it to remove pieces of dirt and other debris. Take the screen out very carefully, remembering how it went into the nozzle. If you put it back the wrong way, the sprayer may not work correctly. You may need to replace a nozzle after several years of use.

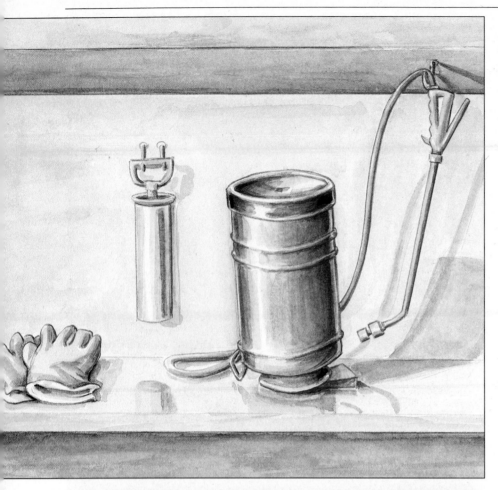

Above: Store your tank sprayer upside down, either hanging or tipped with a wedge, so the inside can dry. Hang the pump beside it, and provide a hook for the wand in order to promote drainage.
Below: Nursery or hardware store staff can help you choose the best product and the right quantity to solve your pest problems.

To clean dusting equipment, tap out excess dust so that humidity doesn't cause it to cake and clog the filter or nozzle. Lubricate any gaskets and the steel rod of plunger-style dusters with graphite rather than oil, because dusts stick to oil.

CHOOSING A PESTICIDE

Your local hardware store or garden center offers a wide array of pesticides to control household pests. When making your choice, be sure you have identified the pest correctly—the pest descriptions in this book should help. Then read the labels and choose a product designed to control that pest.

The product label should also state whether the product is designed to be used indoors or outdoors. Be sure you can use it where you have the problem. For instance, if you want to control ants in the kitchen, be sure that the label states that you can use the pesticide indoors to control ants. If the label says that it is for outdoor use only, choose another pesticide. Formulations intended for use outdoors may stain indoor surfaces or may be too concentrated for indoor use. Each label specifies the product's intended use; never use it for another purpose—that is unwise and illegal.

You can buy many pesticides either in ready-to-use concentrations with applicators or in concentrated form. The choice you make depends on the extent of the pest problem and whether or not you already have your own applicator. If you don't have the equipment to apply the pesticide, you might want to buy a product that does not require a special applicator, such as an ant bait or stake, or one that comes with its own applicator and is ready to use. For large problems it may be more economical in the long run to purchase a good sprayer and use concentrated pesticide. For spot treatments, a ready-to-use formula makes more sense, even if you already have a sprayer.

The sample label (Ortho Insect Spray) shows these headings:

PRECAUTIONARY STATEMENTS
READ ENTIRE LABEL
USE STRICTLY IN ACCORDANCE WITH
LABEL CAUTIONS AND DIRECTIONS.

CAUTION

Hazards

Statement of Practical Treatment

Note to Physicians:

DIRECTIONS FOR USE
How to Use

Plants	Insects	Dosage	Comments

NOTE

Storage and Disposal

Notice:
Chevron Chemical Co© 1987
EPA Registration Number
09876 54321

Chevron
ORTHO
Insect Spray
Makes Up to 8 Gallons Spray

CONTROLS:

ACTIVE INGREDIENTS
INERT INGREDIENTS

Keep out of Reach of Children
CAUTION
NET CONTENTS 8 FL. OZ.

Purchase only the amount of pesticide you think you will use within a year or two. That way you don't have to store pesticides year after year or worry about how to dispose of unneeded products. When you buy only what you need, even the pesticides that have a short shelf life won't have time to decrease in effectiveness.

READ THE LABEL

The most important safety precaution you can take is to read the product label and follow its directions to the letter. Millions of dollars of research went into testing the product to ensure that it is safe and effective when used as directed. That label is your single most important tool in understanding how to use a pesticide. You should read the entire label when you purchase a pesticide, before mixing it, before applying it, before storing it, and before disposing of the empty container. Don't rely on your memory—reread the entire label each time you use the product. Then follow the directions exactly.

All pesticide labels are required by law to list certain key information. Here's what they must contain.
• The name of the product
• An ingredient statement listing all active ingredients and their percentage by weight
• A statement of hazard to people and wildlife, including practical first-aid measures
• The Environmental Protection Agency registration number

• The name of the manufacturer
• Toxicity signal words
• Storage and disposal instructions
• Directions for use, including exact dilution rates and any additional precautions

Signal Words

Pesticides vary widely in how toxic they are to humans, ranging from very toxic to less toxic than table salt. Every pesticide label carries signal words in large type that tell you its approximate toxicity. These signal words are *Caution, Warning,* and *Danger.* All pesticides are potentially dangerous, but the word "danger" on the label indicates the greatest hazard; "warning" means a moderate hazard, and "caution" signifies a slight hazard. Sometimes the concentration of the active ingredients determines the rating. For instance, a ready-to-use formula in a trigger sprayer may rate only a caution, but that same pesticide sold in concentrated form could require a warning.

All labels on pesticide products carry the words "Caution—keep out of reach of children." These words should remind you to treat all pesticides with respect, since even the mildest products, when misused, have the potential to cause harm. Remember that any carelessly handled material is much more hazardous than a properly handled material of higher toxicity.

Study labels before using pesticides and follow instructions exactly. This sample label shows how typical information is arranged under several headings in order to help you locate the details you need.

SAFETY PRECAUTIONS

Pesticides, like medicines, work safely and effectively when used properly but can be dangerous if misused. They can be poisonous if ingested, inhaled, or absorbed through the skin in sufficient quantities. With pesticides, as with all household chemicals, including medicines, cleansers, gasoline, and paint thinner, accidents do happen. Most of these accidents result from carelessness. If you use pesticides according to label directions, you can use them with confidence.

Protective Clothing

It's a good idea to wear protective clothing when you mix and apply a pesticide and when you clean the equipment. As a general safety precaution, cover as much exposed skin as possible. Wear plastic or rubber gloves (not cloth or leather, which can allow liquid to seep through to your skin), long sleeves and long pants, and shoes. If you are going to spray over your head, it's best to wear a plastic hat.

Wear any additional protective clothing indicated by the product label. For instance, a few labels specify that you should wear goggles.

After spraying a pesticide, wash your hands and face with soap and water or take a shower, and put on a complete

When mixing pesticides be sure to use a measuring cup or spoon clearly labeled for pesticide use only, and protect your hands with nonabsorbent rubber or neoprene gloves.

change of clothing. Launder the clothes you wore while applying the pesticide before wearing them again. If the clothing has been contaminated heavily by spills or spray drift, wash it as soon as possible, separately from other laundry.

Proper Mixing

When mixing a concentrated pesticide, do so in a well-ventilated location, rather than in the storage area where you keep your supplies. Open pesticide containers carefully to prevent dust from billowing out or liquids from splashing. Always cut a bag with a knife or scissors, rather than

tearing it open. Be careful not to inhale dusts or sprays at any time as you pour, mix, and apply the pesticide. Pour and measure on a solid, level surface away from areas where food is stored or prepared.

Keep a special set of tools that you use only for mixing and measuring chemicals. Label each tool "for pesticide use only," and store all tools with your chemicals. Never use measuring spoons or cups from the kitchen.

Mix only as much pesticide as you can use right away. Most pesticides don't store well if they are diluted and left even overnight. They begin to lose their effectiveness within a few hours. Measure accurately; too strong a solution may be unsafe, and a weak solution may not be effective.

While spraying, don't leave pesticide containers open or unattended. A curious child or pet may accidentally spill or ingest the chemical. Returning all containers to storage is the safest procedure in most situations.

Application Precautions

Don't eat or smoke while spraying; wait until you've washed your hands and face.

Don't apply pesticides outdoors on a windy day; the spray or dust could blow back onto you or somewhere else you didn't intend it to go. Work indoors only where there is good ventilation but the air is still.

When you are applying pesticides indoors, use only ones that have been formulated for indoor use, as indicated by the label. Do not apply a product in a situation or to a plant that is not listed on the label. In you need to use a pesticide in the kitchen, be sure that the label states that it can be used in this area.

Be careful not to contaminate food, cooking and eating utensils, or dishes. Remove all food and utensils from the area to be treated, or cover them with newspapers before spraying.

If you are treating for crawling insects, such as cockroaches, you needn't spray all surfaces in the infested area. Most indoor crawling pests prefer to crawl along cracks and crevices, such as the bottom edges of baseboards, the back edges of countertops, and cupboards. By spraying only these areas, you actually improve your control by applying the product to targeted areas. You are also using the pesticide judiciously and economically by reducing the amount you apply.

When you spray or dust indoors, avoid inhaling more of the fumes than is necessary. If you are spraying extensively, it's a good idea to leave the windows open so that you have good ventilation. Leave the premises for a few minutes after spraying to give the spray time to settle. If you are spraying an area rather than a surface, close the windows and doors to improve the effectiveness of the spray, but leave the room as soon as you finish spraying. Close the door upon leaving, and wait half an hour or longer before reentering and ventilating the room.

When using a total-release aerosol, shake the container thoroughly, set it to begin spraying, and leave the room immediately. Come back only after all of the contents have been released and they have had time to settle and the surfaces are dry—usually a couple of hours or as directed by the label. When you return, immediately open all doors and windows

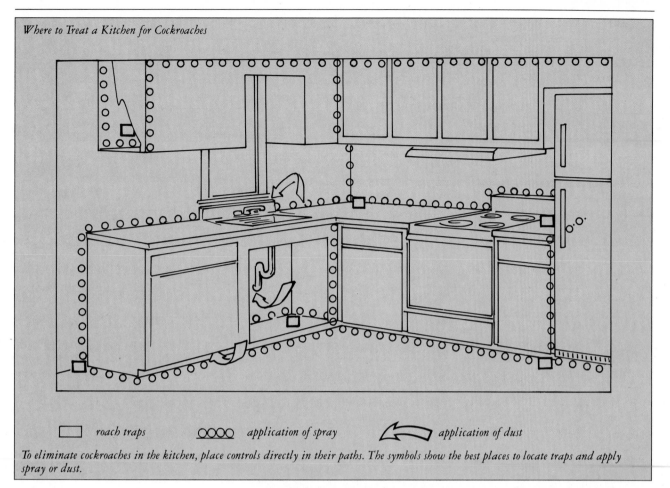

Where to Treat a Kitchen for Cockroaches

| ☐ | *roach traps* | ⵔⵔⵔⵔ | *application of spray* | ⬅ | *application of dust* |

To eliminate cockroaches in the kitchen, place controls directly in their paths. The symbols show the best places to locate traps and apply spray or dust.

to air out the room. If you can still detect an odor, leave the room until it is thoroughly aired.

Do not spray aerosols around fires, pilot lights, or electrical connections. Aerosol products may contain a flammable solvent that can explode if ignited.

Always place baits where children and pets will not have access to them, or choose bait containers designed to prevent children and pets from reaching the bait inside.

When spraying items that may stain, such as rugs and wallpaper, proceed cautiously. Read the product label for any restrictions. Test a small, inconspicuous area first to see if the spray leaves a stain after it dries. If in doubt, use a dust, which won't cause stains.

Safe Storage

Store pesticides in a locked cabinet or locked storage area. The shelves should be strong, stable, and not so high that they can't be reached easily, although

Pesticides and measuring equipment are best stored out of reach of children in a sturdy, locked cabinet. Never store them with any food that you or your pets will eat.

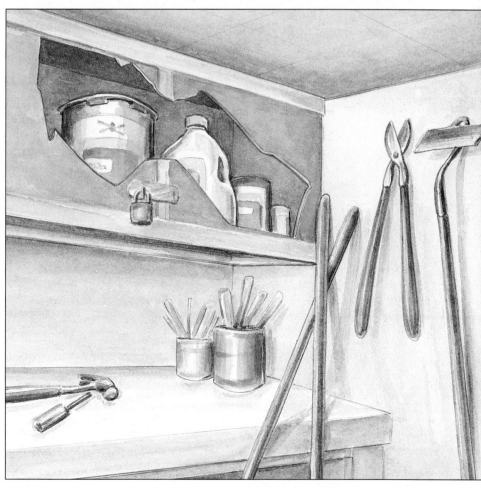

high enough to be out of reach of children. Keep all chemical containers back from the edges of the shelves to decrease the possibility of their falling or being jarred off. The storage area should be well ventilated and away from any food. Avoid storing pesticides near pilot lights or other open flames or sparks, since the fumes could catch fire. Do not store rodent baits near other pesticides because they could absorb pesticide fumes; rodents will not eat baits tainted with the smell of other pesticides.

Always keep pesticides and all other household chemicals in their original containers with their labels attached. Never store pesticides in soft drink bottles or any other containers that could lead anyone, especially a child, to mistake the contents for food or drink. Small children don't know the difference.

Keep herbicides separate from other pesticides because some herbicides are volatile. Their fumes can contaminate other insecticides or fungicides, resulting in plant injury when you apply them to your garden. Check pesticide containers occasionally for leaks, breaks, tears, rust, or similar problems, especially if they have been stored for a long time.

To improve their storage life, store pesticides in a dry, well-ventilated area out of direct sunlight. A garden shed or garage is usually best, but the area should not be subject to temperature extremes.

High humidity, freezing temperatures, or temperatures higher than 110° F can reduce the effectiveness of some products. Tightly close all packages to help keep out moisture and air.

Safe Disposal

If you've mixed more pesticide spray than you need, the best way to dispose of the excess is to spray it according to label directions. Never pour pesticides down the drain—most of them are harmful to the beneficial organisms that work in septic tanks and sewage treatment plants.

When a liquid container is empty, rinse it out with water and spray out the rinse water. After rinsing, replace the lid, wrap the container in several layers of newspaper, tie securely, and discard it in the trash. (Never reuse a chemical container for any purpose.) Do the same with paper pesticide sacks—never burn them. Never puncture an aerosol or pressurized can or put it into a fire; it might explode.

If you have old, unused pesticides that you want to dispose of, diluting them and spraying according to label directions is one of the best ways to get rid of them. Most sprayed materials degrade quickly because ultraviolet light from the sun and microbial organisms help to break them down. If you don't want to spray them, follow the disposal instructions on the label.

IN CASE OF ACCIDENT

If you follow the label directions and take commonsense safety precautions, using pesticides will be as safe as using any other household chemical. Nearly all accidents result from carelessness. In case of an accident with pesticides or any other potentially dangerous household product, be sure you save the container and its label. Most pesticide labels contain an emergency phone number, which can be invaluable in assuring correct and prompt treatment.

If you spill a container of pesticide on the floor, treat the accident with extreme caution. Ventilate the area immediately by opening doors and windows or setting up a fan. Don't light matches nearby because some chemical fumes are flammable. Keep children and pets away by setting up a barrier so that they can't walk in the spill.

Begin cleaning up a spill as soon as you can to minimize your exposure to the product and to reduce possible problems with stains and odors. While cleaning up, wear rubber or neoprene gloves, keep the rest of your body covered, and avoid breathing the vapors. Sweep or shovel spilled solids into a plastic bag, and soak up spilled liquids with an absorbent material such as cat litter, paper towels, or old rags. Place these cleanup materials in a sealed plastic bag, and put it in a trash can outdoors.

In most cases you can wash the remaining pesticide away. Wear rubber gloves and scrub wood, cement, or tile surfaces with a solution of water and strong household detergent. Use a scrub brush or old rags, and repeat the washing until all traces of the chemical are gone. Absorb all cleanup liquids you use with one of the materials just described, place the absorbent materials in plastic bags, and dispose of them.

If you accidentally spill pesticide on your skin, immediately wash the area with soap and water. If you spill concentrated or diluted pesticide on cotton or other porous clothing, immediately change the clothing and wash it.

First Aid

When applying a pesticide, stop what you are doing if you experience any of the following symptoms of accidental poisoning: headache, dizziness, weakness, shaking, nausea, stomach cramps, diarrhea, sweating, or muscle twitching. Many of these symptoms can be caused by other illnesses, such as the flu, but if you have any doubts, get medical attention immediately.

If pesticide gets in someone's eyes, flush the eyes immediately with running water, and continue to do so for at least 15 minutes. Call a physician while this is taking place or immediately afterward for further instructions. Also refer to the product label for instructions.

TROUBLESHOOTING

Occasionally, you may treat an area for pests yet have little success. Many factors influence how effective your treatment is. If you are still troubled by pests after using a pesticide, go over these possible reasons for failure to see what you might have done wrong.

- The pesticide does not kill the pest. Read the label to be sure that the pest is listed on it. If you suspect that the pest might be resistant to the product, choose a pesticide from a different chemical group.
- More pests have migrated from outdoors or from adjacent apartments since the pesticide treatment. Get the cooperation of your neighbors in treating adjacent apartments at the same time. Spray the exterior of the foundation when you spray indoors for pests that may be migrating from outside.

- The application was not thorough enough, allowing many pests to escape the treatment. Be sure to treat all possible hiding and nesting spots at the same time.
- The pesticide you used was not fresh. If stored too long or under improper storage conditions, pesticides will lose their effectiveness. Be sure to use fresh products. Date products when you purchase them, and discard a product after several years if it no longer seems to be effective.
- Alkaline tap water used to dilute the spray caused the spray to break down too quickly. If your water is alkaline, mix sprays with distilled water or purchase ready-to-use formulas.
- The pesticide you are using is at the wrong concentration. Be sure to read label directions and dilute the product correctly.

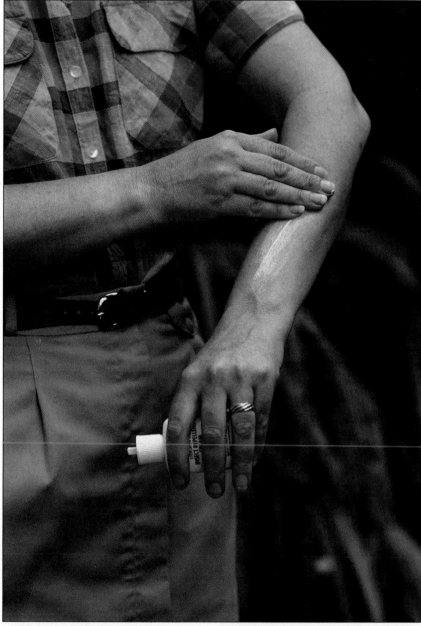

If someone ingests a pesticide, follow the first-aid treatment printed on the container label. While first aid is being administered, call a physician or the nearest poison control center for further instructions. Tell the doctor or poison control center the name of the product and its active ingredients. In the rare instance that you must go to a hospital emergency room, be sure to take the pesticide label with you. If you think that you have been exposed to enough pesticide to make you sick, do not drive a car—have someone else drive you.

OTHER CONTROL METHODS

Pesticides aren't always the answer for controlling every kind of pest. Sometimes a well-aimed flyswatter does the trick. Depending on the circumstances, you may wish to use traps and repellents to control some pests. Combined with properly applied pesticides, these may increase the effectiveness of your control methods.

Repellents

Repellents are products that send signals to pests to keep away. Their odor is disagreeable to the insect or rodent and drives it away. Some repellents work better than others; all must be used properly to be effective.

Repellents applied to your skin and clothing provide an effective way to keep certain biting insects from attacking when you are outside. A number of different compounds are effective and safe to use on skin, but diethyltoluamide, or "deet" for short, is probably the best for most uses. It is also the most common repellent available. Deet doesn't rub or wash off as easily as other repellents do, and it repels a wider variety of pests, including mosquitoes, fleas, biting flies, chiggers, and ticks.

Because biting insects are more attracted to some people than others, the effectiveness of a skin repellent varies from person to person. In general, the more repellent you put on, the longer it is likely to last. Cover all exposed areas of skin, and also apply the repellent to your clothes if you are wearing thin clothing that insects such as mosquitoes and biting flies can bite through. Reapply the repellent every few hours or when it seems to be losing its effectiveness.

Mosquito coils, candles, and torches containing an insect repellent such as citronella are effective if used properly. They work only when the air is calm and when used in sufficient numbers. Mosquitoes will be repelled only in the area covered by the smoke or fumes.

Mothballs, flakes, and cakes repel or control some fabric pests, but they are effective only when used in an airtight container so that the concentration of vapors can build up. They have little repellent effect when used in a closet, unless it is kept tightly sealed. (See page 35 for more details on using these products.)

Night-flying insects are not repelled by yellow light bulbs, but neither are they attracted to them, as they are to most light sources. The white light from normal light bulbs disorients night-flying insects, causing them to fly toward the light source in a circular, ever-tightening pattern. These insects cannot detect light in the yellow spectrum, so yellow bulbs are useful wherever normal white light bulbs attract troublesome insects.

Ultrasonic devices, which cost anywhere from $20 to $300 or more, produce high-frequency sound waves that are generally undetectable by humans. Many of the advertisements for these products claim that they control, repel, confuse, or prevent the reproduction of a wide variety of pests, including cockroaches, mosquitoes, flies, rats, mice, birds, and bats. In numerous university studies, researchers found these devices had absolutely no effect on insects, birds, and bats. Further research is being conducted with rats and mice. Some of the devices do have a weak repellent effect on these rodents, but they have not proven effective as a control measure.

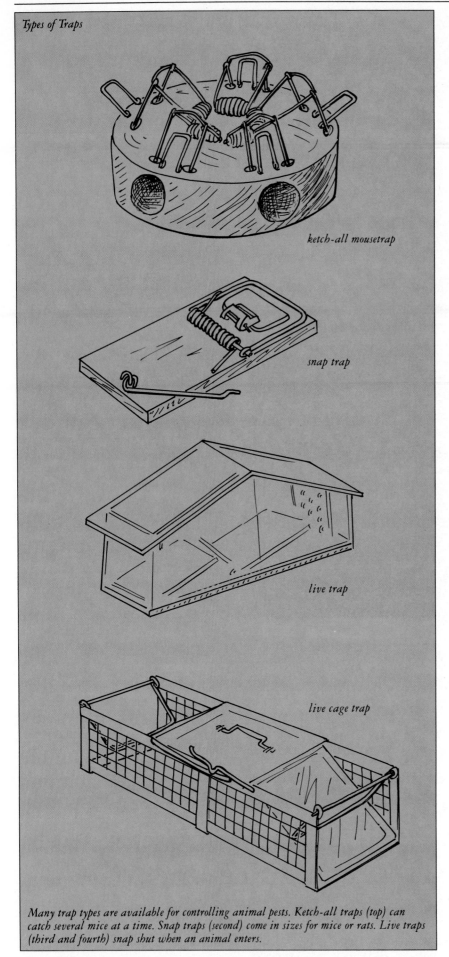

ketch-all mousetrap

snap trap

live trap

live cage trap

Many trap types are available for controlling animal pests. Ketch-all traps (top) can catch several mice at a time. Snap traps (second) come in sizes for mice or rats. Live traps (third and fourth) snap shut when an animal enters.

Traps

Some pests can be controlled with traps, but the trap's effectiveness depends in large part upon how effectively it lures the pests. Once lured, they must be quickly killed or caught.

Light Traps

Also called bug zappers, light traps attract and electrocute night-flying insects. The most effective types use ultraviolet light, often called black light, to attract insects. In general, the higher the wattage of the bulbs, the more insects the device will attract. When an insect touches the grill around the light, it is zapped with anywhere from 500 to 5,000 volts, depending on the model. High-voltage models do not necessarily kill more bugs, but you may not have to clean the electric grill as often.

When used for outdoor insect control, proper placement is crucial or you may lure more insects into your yard than you kill. Always hang the trap behind a wall or thick foliage so that insects in your neighbor's yard can't see it but those on your own property will. Also hang it away from your house, because the dead insect bodies may attract ants and other carnivorous insects.

Light traps attract most species of night-flying pests, including, unfortunately, many beneficial ones, such as lacewings, that devour plant pests. They do not attract most mosquitoes, apparently because mosquitoes are more attracted to heat and carbon dioxide than they are to light.

Mechanical Traps

You can use sticky traps to catch cockroaches and other crawling insects. When it comes to roaches, however, the traps are more useful as an indicator of a pest outbreak than as an effective control method. (See page 39 for more information.)

Fly traps catch flies either with a sticky material (such as flypaper) or by attracting them into a device that resembles an inverted funnel that they can't find their way out of.

Mouse- and rattraps include the basic snap trap as well as sticky traps with a strong adhesive material. A wide variety of new inventions for trapping rodents appears each year—some work better than others. (See page 69 for more information on these traps.)

Traps designed to catch animals alive are useful for catching animals such as squirrels and raccoons. Once you have trapped the animal, you can release it in a

Above: Attractant odors lure flies to a jar where they drown. Such outdoor traps help control flies before they come indoors.
Above right: A properly placed light trap can keep many kinds of flying insects—but unfortunately not mosquitoes—away from your picnic.
Center right: Sticky flypaper lures flies to the trap with chemical attractants. This new model uses drawings of flies as visual attractants, as well.
Below right: Research indicates that this parasitic wasp may have a promising use for controlling household pests. It lays its own eggs in the egg capsules of some cockroach species.

wild area far from your home, but check with public health authorities first. In some areas it may be illegal to release trapped animals because doing so could spread rabies. If a trapped animal seems ill, take it to the public health service. (See page 71 for more information on these traps.)

Predators and Parasites

Encouraging beneficial insects that prey on pests is effective in some situations outdoors, but this technique is not well suited for controlling indoor pests. Most people are very reluctant to allow more insects inside their homes, even if they will prey on the insect pests.

One biological control that has been attempted indoors involves releasing a tiny wasp, *Comperia merceti*, that parasitizes the egg capsules of cockroaches. These wasps are so tiny that most people

never notice them, and they are completely harmless. Unfortunately, they parasitize only the egg cases of brown-banded cockroaches, and repeated releases of the wasps are necessary to provide good control.

Actually, predators may already be at work indoors. Spiders, centipedes, scorpions, and assassin bugs all occasionally wander indoors and are more likely to remain if there are insects for them to catch. Ants are enemies of termites and help reduce termite numbers in some situations, although this is probably rare.

HIRING A PEST CONTROL OPERATOR

You can control most pests yourself, but if you have a particularly severe problem you may want to have a licensed pest control operator do the job for you. Most pest control operators have a thorough knowledge of pest biology and habits; they know exactly what to look for, which pesticide is best, and where to apply it.

In some cases, such as fumigation for termites, the law requires that a pest control operator be the one to make the treatment. Pesticides that require either special equipment or additional training to be applied safely and effectively are placed on a restricted list by the EPA; only a licensed pest control operator who has the necessary equipment and training can buy and use these products.

Choosing a Reliable Firm

Most pest control operators are reliable, respected individuals, and you choose one as you would any other home service provider. Ask friends and neighbors who have used pest control services for a recommendation. You can also check with the Better Business Bureau to see if there are any unresolved complaints against a particular company. It's a good idea to get an opinion and quotes from at least two exterminators, especially if the treatment is costly or if the control techniques sound as if they are of questionable value.

Be especially wary of people who come to your door or phone you without being asked, or who try to scare you into authorizing an immediate treatment. Termites, in particular, may have been in your home for years, so take your time rather than rushing into making a decision—a few more weeks won't matter.

The level of professionalism among pest control operators has increased greatly in the past few decades. They must be licensed, and to become licensed

they must pass tests that demonstrate their knowledge of pests and safe pest control procedures. In addition, many belong to state pest control associations or to the National Pest Control Association. Such a membership means that the company subscribes to a code of ethics, and it also gives the company access to valuable up-to-date technical literature.

Follow-up Measures

Because pest problems are almost always best controlled by a combination of chemical and nonchemical measures, even a professional job of spraying will not control many insects for long if conditions are perfect for them to increase again. For the best control, follow the preventive techniques outlined on pages 6 to 12. If you follow these techniques, you can go much longer between professional spray treatments.

Pest control operators are aware that some people try to rely solely on insecticides to control pests in poor sanitary conditions. Any guarantee they give may require that the buyer of the service meet certain conditions. Some pest control

Above: If your home needs fumigation for termites, law requires that professional pest control operators do the job.
Above opposite: Professional termite control operators will drill through concrete slabs to inject insecticide into the soil on both sides of the foundation.
Below opposite: As a precautionary measure these pest control operators are treating the soil where a house will be built in order to ward off a potential subterranean termite infestation.

operators automatically shorten the terms of a guarantee or charge more if they believe that conditions favor the pests.

Although it may take some vigilance, your home needn't harbor destructive or disease-spreading pests. Remember that effective pest prevention means denying pests food, shelter, and entrance to your home. And when pests do invade, effective control relies on prevention measures combined with the wise use of pesticides, baits, and traps.

KITCHEN AND HOUSEHOLD PESTS

Sawtoothed grain beetle

These most unwelcome of houseguests damage your food, clothing, and peace of mind. Some threaten your health as well. Here you will learn how to pull up the welcome mat and how to send pests packing once they have arrived.

Most homemakers take pride in the cleanliness of their homes and view household and pantry pests as a personal affront. Our homes are our private domains, and we don't take kindly to uninvited guests, especially if they happen to be cockroaches, ants, spiders, or other pests. No one wants to pour cereal into a bowl and find it alive with bugs or don a favorite winter sweater only to discover it riddled with moth holes. But such invasions of household food and effects occur in every household from time to time; your home occasionally may become infested with pests no matter how high your cleanliness and housekeeping standards.

Kitchen and household pests are nothing to be ashamed of, although embarrassment is one of the main reasons why many people attack them with a vengeance. A more important reason to eradicate these pests is that they can spoil your food, damage your clothing and fabrics, and even spread disease. Cockroaches and flies can contaminate food with bacteria that cause food poisoning. And house dust filled with pieces of dead cockroaches may cause many allergy-prone people to develop asthma. Crickets, clothes moths, and carpet beetles ruin clothing, blankets, upholstery, and carpets by riddling the fabric with holes.

No home can be made completely pestproof. Even the cleanest home seems to have enough food and water to feed scavenging insects such as cockroaches and ants. And some household pests, such as boxelder bugs, elm leaf beetles, and clover mites, search not for food and

Shut out pantry pests with a clean and bug-tight pantry. Tightly sealed glass, metal, and hard plastic containers are the best defense.

water but simply for a warm, cozy spot to spend the winter. Others, such as ants, sometimes gather indoors temporarily when it is too dry, too wet, too hot, or too cold outdoors. Still other insects, such as field crickets and many flies, wander in accidentally and would just as soon be outdoors.

Homes tend to be too dry for many insects, and they may not provide the kind of food certain insects need. If such insects wander indoors, they may be annoying or frightening, but do no real damage. Unless they are able to discover a way back outdoors, insects such as boxelder bugs won't live long. Other pests, such as spiders and ants, may even serve a useful purpose outdoors in helping to control other insects or acting as nature's cleanup crew. Their status tends to change quickly, however, when you find them indoors.

Many household pests are very effective at making your home their home. They can find places to hide where it may never occur to you to search. It often takes a carefully laid strategy to get rid of them. The first step is to make food and water unavailable to pests. This may mean storing food in tightly sealed glass, plastic, or metal containers since food pests can easily enter cardboard or paper packages. The second step is to discourage pests from entering your home; many household invaders, such as ants and crickets, begin their infestations outside. Sealing up cracks around doors and windows and keeping screens in good repair helps keep many of these pests outside.

The final step in dispatching these household invaders successfully is to properly apply the right pesticide. Be sure to carefully follow the safety instructions described in "Safe and Effective Home Pest Control," because you may be applying these chemicals in your kitchen, where safety precautions are especially important. Many of the pests included in this chapter, such as cockroaches, ants, flies, and crickets, can be controlled with poisoned baits or with traps. Others may best be controlled with an insecticide sprayed around the house's foundation to eradicate them before they enter your home.

ANTS

At one time or another, almost everyone has seen lines of ants marching across a kitchen counter, pantry shelf, or up the side of a wall. Ants usually don't damage anything indoors, nor do they transmit disease. Nevertheless, if they get into stored food, throw it out, or at least discard the part that was exposed to the ants.

Most ant species, except for fire ants and harvester ants, don't bite or sting. (See page 74 for information on biting ants.) Carpenter ants, which are much larger than

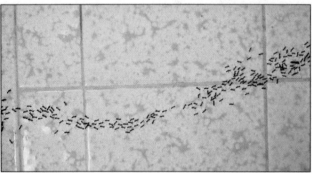

Scout ants mark the way to spilled food by leaving a scent trail. Other worker ants follow the trail and then carry the food back to the nest.

most ants you find around your home, are the only ones that damage wood. They don't eat the wood but simply excavate it in order to create nesting places. (See page 54 for more information about carpenter ants.)

Outdoors, ants eat seeds and dead insects; many ants favor honeydew, a sweet substance that aphids and other insect pests excrete as they suck juices from plants. Ants invade your home in search of food when their outdoor sources are in short supply or if the ant colony is very large. Depending on the species, worker ants may go after sweets, starches, fats, grains, or all of these foods. During dry periods they may come indoors to look for water. When a scout ant finds a food source, it lays down a scent, or trail pheromone, as it travels back to the colony. Other

ants soon find the scented trail and follow it to the food.

The Pest

Ant colonies are highly structured societies consisting of several different kinds of members, each with assigned duties. New colonies begin when winged females and winged males develop in mature colonies. These mate and fly off to find a site for a new nest. There they rub off their wings, but the male dies soon after mating, leaving the female to begin the colony.

In the beginning the female, or queen, divides its time between looking for food, laying eggs, and feeding and caring for the young ant larvae. The first few ants to mature are small workers—sterile females—that help the queen gather food and care for the young. Gradually, the workers begin to do more of the daily chores, and the queen concentrates solely on laying more eggs. As the colony grows larger, some species of ant produce large workers, or soldiers, that are better equipped to defend the colony or to overcome small prey.

Only mature colonies produce winged, reproductive males and females. These often leave the colony in large numbers in spring or fall, but only a very small percentage of females successfully establish new colonies. More often than not, birds and other predators eat them or they are not able to find a suitable nesting site or enough food.

The reproductive females of some ant species often remain with the original colony, eventually producing as many eggs as the queen. Multiple-queen colonies have tremendous reproductive potential and grow quickly if food is plentiful. Portions of these colonies may later split into new colonies.

Most ants build their colonies, or nests, outdoors in open ground, under stones, concrete, or other objects, or in decaying tree stumps. You see only the foraging workers indoors. However, some ants nest indoors or move their nests indoors temporarily, particularly if it gets too wet or too hot and dry outdoors.

You may mistake ants for termites, or vice versa. Take a look at the drawing on page 59 if you have any doubt about which insect you have. Ants have elbowed antennae and thin waists.

Prevention and Control

Whenever possible, find and treat the ant nest, rather than the ant trail, with an insecticide containing diazinon, Dursban® (chlorpyrifos—Ortho Home Pest Insect Control), or Baygon® (propoxur—Ortho Ant, Roach & Spider Killer). Spraying only the ants you see kills but a fraction of the entire colony and does not kill the queens, which continue to lay eggs.

If you can't find the nest, treat the trail; places where the ants may enter the building, such as around door and window frames; and along baseboards. If ants are a serious problem, consider treating around the entire foundation and along concrete patios and walkways. Use a spray or granules containing diazinon (Ortho Diazinon Granules or Ortho Diazinon Insect Spray), Baygon® (propoxur—Ortho Ant, Roach & Spider Killer), Dursban® (chlorpyrifos—Ortho Home Pest Insect Control or Ortho-Klor® Indoor & Outdoor Insect Killer), or Sevin® (carbaryl).

Use these same insecticides to treat any anthills you find outdoors, especially if they're the source of your indoor problem. Apply the insecticide to the hills as well as to the surrounding soil.

One of the best ways to control ants is with an ant bait (Ortho Ant Killer Bait). Baits are sold in stakes, tablets, cups, and bottles. The ants eat the poisoned bait and carry it back to the colony, where they share the deadly snack with other workers, the queen, and the larvae before it kills them. You may be able to eliminate an entire ant colony with an ant bait.

Control ants crawling on outdoor plants by spraying or dusting the ground beneath the plants. This kills the ants as they crawl over the treated ground, yet doesn't harm the

Ant Baits

Ant baits are especially useful when you can't locate the ant nest or if it is difficult to treat. You can use these poisoned cups both indoors and outdoors. Place them along indoor ant trails or the house foundation in spots where pets and children can't get at them. Because ant bait must compete with other food the ants like, such as honeydew and dead insects, it works best from mid-fall through early spring, when such food is in short supply.

Baits frequently do such a good job of control that you won't have any ant problems for a long time afterward. Some baits, however, don't work on all ant species. Species that feed entirely on greasy foods or seeds ignore sugar-based bait. In such cases, choose a bait that is effective against grease-eating ants, such as Ortho Ant Killer Bait. You also must be patient when using a bait, because it may take several weeks to eliminate the entire colony.

predators and parasites that help control honeydew-producing insects. Use diazinon (Ortho Diazinon Insect Spray), Sevin® (carbaryl—Ortho Sevin® Garden Spray), malathion (Ortho Malathion 50 Insect Spray), or Dursban® (chlorpyrifos—Ortho-Klor® Indoor & Outdoor Insect Killer). Spraying the plant directly to control the sucking insect pests also controls ants on the plant; be sure that the infested plant is listed on the product label.

One way to control ants nesting in the soil of potted plants is by gradually immersing the container in water. The ants move higher in the soil as their lower galleries flood. For medium-sized containers, it may take only half an hour to flood out the entire colony. Leaving the roots immersed for longer than necessary may damage them.

Ants often disappear when you keep spills of liquids and particles of food cleaned up. Don't tempt them with dirty dishes or open containers of stored food. Sealing up cracks inside and outside your home where ants may enter usually won't keep them out unless you combine this with good sanitation and insecticides.

BOXELDER BUGS

Large numbers of boxelder bugs may invade a home or garage in search of a winter hiding place. They don't feed on any household items, but their excrement can leave spots on drapes, furniture, and clothes. If you accidentally crush one of these bugs, it produces a strong, disagreeable odor. Although boxelder bugs can bite people, they rarely do so.

The Pest
Adult boxelder bugs are about ½ inch long and a third as wide. They are dark brown to black with thick, bright reddish orange lines on their backs and wings.

Controlling boxelder bugs outdoors and sealing entry spots will limit their annoying habit of moving indoors in the autumn.

In late summer or early fall, the adult bugs begin searching for a dry, sheltered place to spend the winter. Although they can fly 2 miles or more, boxelder bugs usually crawl from place to place. They overwinter close to their host trees in hollow tree trunks, cracks and crevices around foundations and windows, inside sheds and homes, and in similar places.

You may not notice boxelder bugs inside until a warm winter day, when they become active and scatter through the house. You may find hundreds of these bugs gathered on a sunny porch or south- or west-facing wall. During these periods of winter activity, more boxelder bugs may find their way indoors.

In early to mid-spring the bugs return to their host trees. After feeding for several weeks, the females lay eggs on or near the tree, especially in bark crevices. Nymphs, which resemble their parents but are bright red, later turning darker red, hatch in two weeks. They mature in about two months, completing two generations a year in warmer areas of the country and one generation in cooler areas.

Boxelder bugs eat primarily seedpods of box elder trees, both while the seeds are developing and after they fall to the ground. To a lesser extent, they feed on tender foliage and young twigs. They generally attack female box elder trees. These pests may also infest ash and maple trees, each of which has seedpods similar to those of the box elder.

Prevention and Control
You can vacuum these annoying pests or spray them with a product containing pyrethrins. Because more bugs may enter your home on the next warm day, it is a good idea to control boxelder bugs outdoors before they enter. Do so by spraying along the foundation of your house, at the bases of host trees, around woodpiles, and in similar places where the bugs might be gathered. Use diazinon (Ortho Diazinon Insect Spray), applying enough spray to thoroughly wet all surfaces where the bugs crawl.

Reduce the number of bugs coming indoors by repairing and sealing up entry spots (see pages 7 to 11 for details). Rake up leaves, weeds, and grass from a 6- to 10-foot-wide strip around the house foundation, particularly on the south and west sides. This tends to discourage bugs from congregating near the foundation.

Spraying infested trees in summer and fall prevents the insects from becoming an indoor problem later. Apply malathion (Ortho Malathion 50 Insect Spray) to the trunk and foliage of the infested tree according to label directions.

CARPET BEETLES AND CLOTHES MOTHS

The larvae of carpet beetles and clothes moths damage clothing, upholstery, and carpets more often than any other insect pest. Their adult forms do no harm. Although you would think from their names that the beetles primarily eat carpets and the moths mainly eat clothes, this is not the case. Except for minor differences, they both eat the same things, and control measures are identical.

Other insects that damage fabrics, but do so less often, include silverfish, cockroaches, crickets, and termites. You can best identify the source of damage by seeing the pest.

The Pests
Adult carpet beetles are similar in shape but slightly smaller than ladybugs. They may be either totally black or mottled brown and yellow. You can sometimes observe carpet beetles outdoors feeding on flower pollen. Since they fly to lights, you may see them

Bristly larvae feed on carpets, upholstery, and clothing for up to three years before hatching into speckled adult carpet beetles.

on windowsills indoors. The adult form of this pest does not eat fabrics, but if you see the beetles, you'd best search for their destructive larvae.

Carpet beetle larvae are brown, about ¼ inch long, and covered with bristles. They feed for between nine months and three years before

In order to be effective, mothball fumes have to build up in a sealed container; an airtight box, trunk, or plastic bag makes a good choice. Wrap mothballs in paper before inserting them in clothing, making sure that they don't touch the fabric or plastic.

they pupate into adult beetles. Because they crawl around quite a bit, you may notice them on walls or on fabrics that they aren't actually infesting. They leave brown, bristly cast skins near the damage site, giving a clue to their presence.

Adult clothes moths are small tan to yellowish moths with a wingspread of just under ½ inch. The moths never eat throughout their 20- to 30-day life—they simply mate, disperse, and lay eggs. They tend to stay in darker areas of rooms, and are not attracted by lights. If you see moths flying around lights at night, be assured that they are probably not a fabric-damaging species. Such moths either flew in from outdoors or are food-infesting moths.

Clothes moth larvae are whitish with black heads and grow to nearly ½ inch long. The larvae of the casemaking clothes moth enclose themselves in silken cases, which they drag around wherever they go. The larvae of the webbing clothes moth may crawl around without cases, but they usually feed from within silken tunnels that they build in the fabric. You may not notice these tunnels because the larvae incorporate strands of the fabric into the tunnel, thus camouflaging it. Clothes moth larvae may feed

for as little as 35 days or for as long as 2 years or more before they pupate.

Carpet beetles and clothes moths prefer items of animal origin, including wool, hair, fur, horns, and feathers. If something of nutritional value to the insects, such as perspiration, urine, beer, or fruit juice, soils the fabric, the pests consume the stained area first. In fact, clothes moths do not develop properly on wool that is completely clean. They also eat cotton and other items made from plant materials. They'll eat synthetic fabrics, but only if the item is soiled or stained. Occasionally, these beetles and moths get into spices and stored foods.

The larvae of both carpet beetles and clothes moths shy away from light and do their worst feeding damage in dark, undisturbed locations, such as under furniture, in drawers, and behind baseboards and heaters. In carpeting, they usually feed well down in the pile or between the carpet and pad.

Carpet beetles and clothes moths enter your home in a couple ways. You may inadvertently bring an infested item indoors, not noticing adults or larvae of either

You may have to look closely to notice the feeding tunnels made by webbing clothes moth larvae as they devour fabric.

insect hidden in fabric folds, in cracks of an old clothes chest, or in some other item.

The adults of both insects can also fly in. Nests of birds, rodents, bees, and wasps

around the outside of your house provide continual sources of infestation. Carpet beetles and clothes moths often feed on animal hair and feathers in such nests. When it gets cold outdoors, or when the food in these nests begins to run out, the insects may try to fly or crawl indoors.

Prevention and Control

Effective control means that you must be a very thorough detective to find every place the insects are breeding. Keep in mind when you find and eliminate a breeding site that adults or larvae may have already left it and begun infestations elsewhere.

The places in which these pests are most commonly found are in clothes, carpets (especially in dark areas under furniture), blankets, woolen rugs, upholstered furniture (between the cushions, on the underside of the item, and in other less disturbed areas), felts in pianos, animal bristles in brushes, feathers in hats, animal skins, and accumulations of lint or pet or human hair behind baseboards, in heaters, under furniture, or in cracks and corners.

When you examine clothes, look most closely at woolen items in storage, being sure not to neglect hidden areas of the garment, such as under the collar or inside the pant cuffs. Make a thorough search of your pantry, paying particular attention to such items as cereal, dog food, fish meal, and old spices.

Outdoors, check to see if there are bird, bee, or wasp nests next to your home, and remove them. If you find a nest containing baby birds, you can safely wait until after they have grown and left before removing the nest. A dead bird in an unused chimney or a dead mouse inside a wall can also be a source of insects.

Once you find where the insects are breeding, you can deal with them in several ways. Both dry cleaning and laundering kill the insects.

All kinds of organic material—even feathers—are food for clothes moths. Casemaking clothes moth larvae, shown here, carry silken cases with them as they eat.

Many people also get good results by thoroughly brushing the fabrics outdoors and then hanging them in the sun for several hours. Insects that you don't actually brush off fall from the fabric when they cannot find protection from the sun. Cleaning and/or spraying the fabric gives it added protection after sunning.

Infested furs are particularly difficult to treat. Do not place furs in the sun, because they may fade and dry out; do not spray them with an insecticide. Either place furs in a commercial storage facility, where they will receive professional care, or place them in tight containers with para-dichlorobenzene or naphthalene, commonly used in mothballs.

You can spray rugs, furniture, or clothes directly with a formulation of pyrethrins, such as tetramethrin and sumithrin (Ortho Moth-B-Gon® Moth Proofer) or another insecticide labeled for fabrics. Do not use pesticides not labeled for fabric, because they may cause stains. Spray lightly and uniformly until the surface is moist. Spray infants' clothing or blankets only if you plan to store them. Launder before using them again. If you do not want to treat an infested rug or piece of furniture yourself, find a pest control or carpet cleaning firm experienced in treating rugs and carpets.

If you throw out a damaged item, it is a good idea to spray it first with insecticide and put it in a sealed plastic bag so that the insects don't escape and disperse from your garbage can.

Spray with a residual insecticide in closets, chests, or other areas where you have found carpet beetles or clothes moths. Use Baygon® (propoxur), diazinon, or Dursban® (chlorpyrifos—Ortho Home Pest Insect Control). Apply the insecticide along the edges of wall-to-wall carpeting, behind heaters, and in corners, cracks, baseboards, and other hard-to-clean places. In closets, spray

Finding adult clothes moths in your home should warn you of an infestation. The moths flutter about near dark corners; unlike most moths, they are not attracted to light.

all corners and cracks, as well as shelves and the ends of clothes rods. This controls hidden insects and kills others that may crawl over the treated surfaces.

You can help limit or prevent an infestation of clothes moths and carpet beetles in several ways. Vacuum carpets regularly. Occasionally move all low furniture pieces so that you can vacuum the rug underneath. Keep furnace air ducts free of lint and dust. Always empty the cleaner bag promptly after you vacuum an area of suspected infestation, because it may contain eggs, larvae, or adults. Occasionally move around and air out rugs and clothing.

Because fabric pests particularly attack things in storage, proper storage is

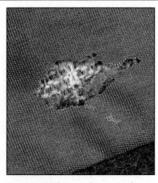

Here a clothes moth larvae has eaten a hole in the outer layer of a garment, but has left the lining alone.

extremely important. Always clean clothing, blankets, and other susceptible items before you store them, and spray with tetramethrin and sumithrin (Ortho Moth-B-Gon® Moth Proofer). For best results, wrap the treated garments in a plastic bag before storing. You may also wish to spray cracks and crevices in the storage closet or chest with the moth-proofing product. Dry-clean or launder garments before wearing.

You can also store woolen items in a tight container with mothballs, moth crystals, or flakes. Contrary to popular belief, the odor of mothballs in a closet is not enough to kill or repel fabric pests. The fumes must build up to a high concentration in a confined area before they are effective. You need a tight container, such as a trunk, chest, box, or thick garment bag. Before you rely on a chest to be airtight, inspect it and tape or caulk any cracks. You can also wrap clothing in heavy paper or plastic and tape it closed along the edges to make an airtight container. Be sure that the mothballs don't actually touch the fabric—separate them with a layer of paper. Use 1 pound of mothballs, moth crystals, or flakes for every 20 cubic feet of storage space. (One-half pound is enough to treat a good-sized trunk.) Air out, launder, or dry-clean garments before wearing.

Cedar chests can make good storage containers primarily because they are tightly constructed, and this keeps the pest from finding its way inside. The cedar oil in the wood may control fabric pests initially, but it loses its potency after a few years. Cedar oil, like mothballs, works only if the space is enclosed; cedar-lined closets do not control insects.

CLOVER MITES

Large numbers of tiny brown or red mites often invade homes from fall through spring. If you crush one, it leaves a red spot that stains whatever it's on. These harmless insects do not bite humans or pets, carry diseases, or feed on stored food or furnishings. But if present in large numbers, they become annoying pests.

The Pest
Clover mites are eight-legged creatures that are more closely related to chiggers, ticks, and spiders than they are to insects. They are brown or reddish and smaller than the head of a pin. You can identify clover mites by looking at their front legs, although you may need a magnifying glass to do so. The front legs are longer than the body and about twice the length of the pests' other legs.

Clover mites lay their bright red eggs in crevices on bark, fence posts, and foundations; behind shingles and siding; and in similar places. The young mites hatch and grow quickly, becoming adults in less than a month. Two main generations mature each year; eggs laid in late spring hatch in the fall, and those laid in the fall hatch early the following spring. In some areas a portion of the eggs laid in the spring hatch in the summer, producing a third generation. Most hatching occurs at between 45° and 75° F.

Clover mites feed on grasses, clover, and a large variety of other garden plants and weeds. Heavy feeding causes leaves to develop a silvered appearance. In the fall, clover mites gather on vegetation around homes and on foundation walls. When cold weather arrives, the mites become abundant along warm concrete foundations, especially on the south sides of homes, where the sun provides the greatest amount of heat. They have a tendency to crawl up the exterior walls until they find a crack to enter. Cracks along foundations, doors, and windows are typical entry points. Many mites end up underneath siding, inside walls, behind baseboards, and in similar protected places. They remain inactive until a warm day, when those already in the walls move further indoors, and additional mites crawl in from outdoors.

This activity and movement may occur periodically during warm spells throughout the winter. In the spring, clover mites become even more numerous indoors, apparently because they crawl

Seeking warmth, clover mites sometimes enter a house in the fall. It is better to vacuum them up than to squash them, since they leave a red stain.

about in search of food and mistakenly wander inside. Homes on new sites or with well-fertilized lawns growing close to their foundations have the most problems with clover mites.

Occasionally, other mites that look like clover mites

become abundant in yards but do not invade houses. Some of these are subspecies of clover mites and are so similar that they can be distinguished only by a specialist who is familiar with this particular group of mites.

Prevention and Control

To avoid leaving red stains, carefully vacuum up live or dead mites indoors without crushing them. If any of the mites are alive, seal the sweeper bag and dispose of it as soon as you finish the vacuuming.

Spraying mites indoors with Dursban® (chlorpyrifos—Ortho Home Pest Insect Control) or diazinon kills them. To provide further relief from mites crawling out of hiding places, spray along baseboards and doors, between windows and screens, and in similar places.

You gain only temporary control of clover mites indoors if you do not take steps to control them outdoors around the foundation. Treat a band 15 feet wide around the foundation by spraying with diazinon (Ortho Diazinon Insect Spray), malathion (Ortho Malathion 50 Insect Spray), Dursban® (chlorpyrifos—Ortho Outdoor Ant, Flea & Cricket Spray), Baygon® (propoxur—Ortho Hornet & Wasp Killer), or dicofol. Also, thoroughly treat the base of the foundation and places where mites could enter your home or lay eggs, such as places where siding overlaps the foundation, vents, and window and door frames. Treat again if you have a heavy rain within a few days of the first treatment.

In severe infestations, make a second treatment 7 to 10 days after the first one. If you find large numbers of clover mites in the lawn, treat it as well. Where clover mites are a recurring problem, spray in the early spring and again in the fall before they begin

to come indoors. Creating a 18- to 24-inch-wide strip of loose soil around the foundation provides an effective barrier to clover mites not already on the foundation. Spade or otherwise cultivate the soil to create this barrier, and keep it free from grasses and weeds. You can use the strip by itself or to supplement spraying around the foundation. If you use it as a supplement, make the barrier before you apply the insecticide so that you leave the layer of insecticide intact.

Sealing cracks along doors, windows, and other places where mites may enter aids control but is not very effective, because clover mites' small size allows them to enter very narrow cracks.

COCKROACHES

Cockroaches are probably the most repulsive of all household pests. Householders hate them so much that they spend $2 billion every year to battle these pests. Roaches infest food but also eat the glue from stamps and labels on canned goods and the starch in bookbindings. Cockroaches will also eat leather

The smallest and most common cockroach species, the German cockroach is the only one that carries its egg cases until they are about to hatch.

Cockroaches are not only a nuisance and an embarrassment, they are suspected of carrying and transmitting a wide variety of disease organisms. Although it is very difficult to prove conclusively that cockroaches spread germs, many authorities believe that they do. Salmonella bacteria, a major cause of food poisoning, may be carried by roaches. They may also spread parasites that cause toxoplasmosis, a flulike illness that can cause birth defects. Because cockroaches visit very dirty places, they come into contact with a wide variety of disease organisms. These can adhere to their bodies or be ingested and spread through their feces.

American cockroaches are the largest of the roaches that live in the United States; they sometimes reach 2 inches in length.

and stained clothing. You may find them nesting during the day in folded paper bags or even in a desk drawer. If present in large numbers, these pests give a room a foul odor. They emerge at night to feed and will scatter rapidly if you turn on the lights.

The insects spread these germs by dropping their feces on food or dishes or simply by walking across a surface.

A recent finding indicates that certain people may have an allergic reaction to house dust or food containing pieces of dead cockroaches or

their feces. The most susceptible people are those prone to other allergies and who live in places heavily infested with cockroaches. These allergies can be so severe as to cause asthma and may explain the large number of asthma sufferers found in city slums.

Cockroaches can bite people, but this very seldom happens. The species most likely to bite are the large outdoor species, particularly if they become very numerous. These pests have been found at night gnawing on fingernails, eyelashes, callused portions of hands or feet, and open wounds. Fortunately, such occurrences are rare.

Cockroaches enter homes by various means. They may be brought in from infested areas in grocery bags, produce, laundry, kitchen appliances, and furniture. In apartments, openings around water pipes and electrical lines are just two of the "highways" cockroaches use to travel between units. The species that can also live outdoors in warmer climates can crawl into homes through any available opening, including any cracks around doors and windows.

Most cockroach species originated in Africa and were accidentally spread throughout the world on ships via major trade routes. They adapt themselves so well to human-made structures that today cockroaches can be found in such unlikely places as Arctic outposts and offshore oil platforms.

The Pest

Several species of cockroach—German, American, brown-banded, oriental, and smokybrown—infest buildings. Asian cockroaches, an aggressive species, were recently introduced to Florida. (Their common names do not indicate their origins.) Each species looks and behaves slightly differently, but they all have certain things in common. The females produce egg capsules, each of which contains an average of from 14 eggs (oriental cockroaches) to as many as 40 eggs (German cockroaches). The wingless young, or nymphs, are colored slightly differently than the adults. The nymphs shed their skins (molt) several times as they grow larger. The last time they molt, they emerge as winged adults, although most cockroaches cannot fly or can fly only very short distances. Asian cockroaches are strong fliers, however. The entire life cycle may take as little as 2 months or as long as 2½ years, depending on the environmental conditions and species.

Cockroaches are active at night and hide during the day in sheltered, dark, tight places, especially places close to food and water. Likely places for roaches to be found indoors include under the kitchen sink; in the toilet tank; under, behind, or inside kitchen cabinets and drawers; behind loose baseboards; under or behind appliances; inside folded paper bags; inside desk or dresser drawers; in stacked newspapers and magazines; and similar sites.

Cockroaches found indoors are likely to be one of the following species. Proper identification will help you understand how best to control the particular cockroaches infesting your home.

German cockroaches, the most common species, are also among the smallest. The adults are only about ½ inch

Slower moving than other common species, the oriental cockroach prefers cool, damp areas and can live outdoors.

long, light-brown, and have two dark brown bands behind their heads. German cockroaches live all over the house but are particularly common in kitchens and bathrooms.

German cockroaches are the only common roaches that carry their egg cases with them until they are almost ready to hatch. They lay more eggs (averaging 7 to 8 egg cases, each containing 35 to 40 eggs, for a total of 320 eggs per female) than other cockroaches. They also develop more quickly, completing as many as four generations a year. Under favorable conditions, a female can leave more than 30,000 descendants in one year.

Brown-banded cockroaches are about the same size as German cockroaches and are also light-brown, but they lack the two bands near their heads. Females and nymphs have distinct brown bands that circle their abdomens. These roaches can tolerate warmer, drier conditions than German cockroaches can; they may be found throughout the home, even in fairly warm, dry areas. Brown-banded cockroaches carry their egg cases with them for a day or two before depositing them, often in clusters, in protected places. These can be found on the undersides of tables and furniture, in the corners of cabinets, and other dark areas.

Brown-banded cockroaches can complete slightly more than two generations a

year if the temperature is higher than 80° F. Because most homes are kept cooler, they may take considerably longer. Under optimum conditions, a female of this species could leave more than 600 descendants in a year.

Adult oriental cockroaches are from 1 to 1¼ inches long and dark brown to black. They move rather sluggishly compared to other cockroaches, even when disturbed. They live in cool, damp areas fairly close to moisture, such as in sewers, in water meter boxes, under porches, and in basements. Oriental roaches stay mostly in basements and the ground floors of buildings. During warm, humid weather they may live outdoors in ground covers, compost piles, and similar areas.

The females carry their egg cases for only a short period before dropping them or attaching them to protected

surfaces. These cockroaches reproduce comparatively slowly; each female can leave only 200 offspring in a year. It may take a full year or longer to complete a life cycle.

American cockroaches are the largest household cockroaches and are 1½ to 2 inches long as adults and reddish brown to dark brown with a yellowish marking behind their heads. They are quicker than oriental cockroaches but have similar habits. You will find them in moist places, usually below the second story of a building. They also live outdoors in warm climates, in areas where they are called palmetto bugs or waterbugs.

Like oriental cockroaches, female American cockroaches drop their egg cases or glue them to a protected surface a few days after they form. Populations of American cockroaches can build up more rapidly than oriental cockroach populations; a generation may be completed in about eight months, although it frequently takes longer. Each female can leave over 800 offspring in a year. They can live outdoors in warm climates.

Smokybrown cockroaches are 1¼ to 1½ inches long and darker than American cockroaches. These cockroaches commonly live outdoors in wood and compost piles, leaf litter, ground covers, and similar areas in warm climates. They tend to become more of a pest when the weather turns either cold or hot and dry, which drives them indoors.

Brown-banded cockroaches don't require much moisture, so may be found throughout the house and even in office buildings.

The females carry their egg cases for a day to two before attaching them to protected places. Outdoors, they are most likely to leave them

Smokybrown cockroaches are often found outdoors, where they inhabit woodpiles and leaf litter.

around wood and compost piles, in crevices in concrete foundations, under wood siding and eaves, and around window frames and weep holes. They can complete a generation in about eight months. Each female can leave about 300 offspring in a year.

Besides these more common cockroaches, there are many other species that live outdoors in warm climates. Called wood roaches or field roaches, these stray indoors from time to time but almost never become as abundant as the household cockroaches. A potentially serious new pest, the Asian cockroach, presently occurs only in central Florida. It is close kin to the German cockroach, but it is very bold and is active in the early evening. Because it can fly and is equally content living in shaded areas of leaf litter and thick grass as it is inside houses, the Asian cockroach may become a serious and more widespread pest. Researchers are concerned that this pest could accidentally be transported in cars and trucks and spread throughout the southern states and up both coasts.

Prevention and Control

Planning and diligence are required if you want to do a good job of cockroach control. Spraying or stomping on only the roaches you see never really does much to control an entire population. For every roach you see, scientists

estimate that, depending on the species, there may be 100 to 600 more in hiding.

To control indoor cockroaches successfully, you need to do three things: (1) reduce or eliminate their food, water, and shelter; (2) eliminate the present roach population, usually with a spray, powder, or bait; (3) prevent additional cockroaches from getting into your home. You can't expect to get good control for long if you don't follow all three of these steps.

Good sanitation, which keeps water and food from roaches, will not by itself eliminate the pests—they can, after all, subsist on glue from postage stamps. Sanitation does, however, greatly improve the control you get with insecticides. In some cases you cannot adequately control cockroaches unless you practice good sanitation. This step is so important that most pest control operators charge more to control roaches where conditions are unsanitary.

Don't concentrate solely on eliminating food sources; water is just as important to roaches. Most species will dehydrate and die within a few weeks if they can't find moisture. They can go for one or two weeks without food. To deprive roaches of their primary food and water sources, be diligent in your housekeeping. Every day, wipe up all crumbs and spills from counters, keep dishes washed, and empty all water from the dish drainer. Wring out dishcloths and sponges. Keep

food tightly covered in sealed containers or in the refrigerator. Keep a tight lid on your kitchen waste can or empty it each evening. Repair any leaks in plumbing. Rinse and dry empty beverage containers. Don't leave pet food out overnight in open containers. Regularly empty the drip pan under a frost-free refrigerator if it collects water.

Eliminating places where cockroaches can hide is another important aspect of sanitation. They must have dark, undisturbed places in which to spend the day. Caulk any cracks in walls behind baseboards and around pipes to reduce the number of hiding places. Don't let newspapers, magazines, and other papers pile up.

A variety of insecticides are available to kill cockroaches, including sprays, dusts, aerosols, and baits. No matter what type you choose, if you live in an apartment it is best for everyone in the building to treat at the same time, or at least on the same day, as you do. Some insecticides strongly repel cockroaches that are not close enough to the site to be killed. If you spray, and your neighbors spray a week later, roaches that aren't killed immediately will simply move back and forth in order to be in the most friendly environment.

Insecticide sprays are used for cockroach control more than any other method. To be effective, they must be applied to every cockroach hiding place. Concentrate your spray treatment on cracks, crevices, corners, and edges where cockroaches most commonly hide. Do not spray entire surfaces—that only wastes the insecticide. Before spraying inside kitchen cabinets or drawers, remove all food, dishes, and utensils, and cover them with newspapers or some other cover. Then spray only along the edges and corners. Wait until the spray dries thoroughly before

returning dishes and utensils. Carefully follow all other label directions.

Many different insecticides are used to control cockroaches. These include Dursban® (chlorpyrifos— Ortho Home Pest Insect Control or Ortho Roach, Ant & Spider Killer), diazinon, and propoxur (Ortho Ant, Roach & Spider Killer). From several days to two weeks these continue killing roaches that walk over the sprayed surface. D-trans allethrin (Ortho Professional Strength Flying & Crawling Insect Killer), tetramethrin and sumithrin (Ortho Household Insect Killer Formula II), and pyrethrins kill roaches more quickly but become inactive several hours after spraying.

A total-release aerosol, such as Ortho Hi-Power Indoor Insect Fogger, will quickly control cockroaches present in exposed locations. The disadvantage is that foggers do not penetrate cracks and crevices where many cockroaches hide. Foggers do, however, flush roaches out of hiding and, if used in conjunction with a spray applied to baseboards and other hiding places, provide excellent control.

Boric acid powder (Ortho Roach Killer Powder) is highly effective when properly applied, and it lasts for months. It does not provide immediate control, however. A week or so may pass before you notice a reduction in the number of cockroaches. The powder clings to the roaches' legs and antennae as they walk over it, and they ingest it later when they clean themselves. Boric acid is fairly safe to handle. It acts primarily as a stomach poison for both insects and people, so never swallow it.

Apply boric acid powder with a spoon or puff duster according to label directions. Treat places where cockroaches walk or hide, such as under and behind kitchen appliances and cabinets, under sinks, behind baseboards, and in other cracks and hiding places. Be very thorough in treating all possible areas where cockroaches might hide. A light dusting in each area is usually more effective than applying a large amount, because thick layers tend to cake and are then less likely to cling to the insects. After treatment, no powder should be visible; sweep any you see into cracks and crevices.

Other dusts and powders for cockroaches can be used in the same way but may not be as long lasting as boric acid powder.

Baits available for cockroaches contain Baygon® (propoxur—Ortho Pest-B-Gon® Roach Bait or Ortho Roach Killer Bait), Dursban® (chlorpyrifos), or boric acid powder. To control roaches outdoors, use a bait containing Baygon® (propoxur—Ortho Earwig, Roach & Sowbug Bait) around the house foundation according to label directions.

Baits are effective in areas where you aren't able to spray or as a supplement to sprays. Because the roaches must both find the bait and feed on

articularly useful for monitoring roach infestations. A few traps placed in key locations will alert you when it is time to reapply pesticide. Because cockroaches reproduce so rapidly, traps cannot significantly decrease a population, even if they seem to catch many roaches. Traps alone can be effective only against the larger, slower-to-reproduce species and in situations where there is no source of reinfestation.

The effectiveness of baits and traps depends in part on proper placement. Place them where cockroaches are most likely to crawl when going to and from food and hiding places. The corners of cupboards and counters and along walls are the best places because cockroaches tend to stay next to objects rather than travel in the open. Place all baits where children and pets can't get to them. The more traps you use the better. Replace them when they become filled up or dirty and are no longer able to hold roaches.

Once you have controlled cockroaches, you can help

baseboards and door frames, as well as cracks in concrete foundations and brick walls and ones where water and sewer pipes and electrical, television, and telephone lines come inside. Securely screen weep holes and ventilation openings.

If cockroaches are coming in from outside the home, remove breeding areas next to the foundation to help prevent a buildup of roaches there. Ground covers, mulches, and compost and firewood piles should be removed to at least 5 feet from the foundation. If roaches continue to find their way indoors, apply an insecticide around the foundation and around garbage cans and other breeding places. Use diazinon (Ortho Diazinon Insect Spray), malathion (Ortho Malathion 50 Insect Spray), or Dursban® (chlorpyrifos—Ortho Home Pest Insect Control).

Because roaches often hitchhike on items you bring into your home, look over anything you bring indoors, especially if it comes from a place where there may be roaches. Beverage cartons, boxes, dried pet foods, grocery bags, used appliances and furniture, laundry, used clothes, and, in warmer climates, firewood are some of the more common items on which cockroaches may be stowed away.

Although some roach populations have become resistant to certain insecticides, if an insecticide fails to control them it is more often than not because you missed a few hiding places or because of poor sanitation. Do not expect instant results, and never increase the concentration of a spray. Follow label directions exactly. If cockroaches increase again after several weeks or months, make another treatment at that time. Try to discover whether you have missed areas that harbor the pests, and practice good sanitation and preventive techniques.

CRICKETS

Although crickets may annoy you with the repetitive loud, chirping sounds that males make to attract females, their worst vice is feeding on household goods. Outdoors they eat plants and dead insects, but indoors crickets eat crumbs and stored food, paper, and a wide variety of synthetic and natural fabrics in clothes, carpets, drapes, and upholstered furniture. Crickets find fabrics stained with food or perspiration especially attractive. They typically chew large holes in fabrics, in contrast to the small holes left by clothes moths.

The Pest

Two crickets, the house cricket and the field cricket, can damage items in your home. The tiny nymph matures into a 3/4- to 1-inch-long adult with two long antennae

Although the field cricket can't breed indoors, it will occasionally stray inside, where it can do as much damage as a house cricket.

on its head and two long appendages on its rear end. The female also has a third rear appendage—a long, swordlike device for laying eggs. The house cricket is light yellowish brown or straw colored and has three dark bands on the back of its head. The field cricket does not have these bands and is dark brown to black. Both species occur throughout the United States, although the field cricket is more common in the western states.

Cockroach eggs can be brought into the house in bags or boxes from infested stores. This German cockroach egg case was deposited in a brown paper bag.

it, baits are most effective when used immediately after you clean up other food sources. Because baits are long lasting, some people put them out after they spray to provide control after the spray becomes inactive.

Cockroach traps, such as Ortho Hi-Power® Waterbug & Roach Trap, contain a bait and a sticky adhesive that traps roaches. These are

prevent their return by sealing up any holes where they may enter. This is particularly important in apartments, where roaches will migrate from next door, and in warm climates, where roaches may be breeding outdoors. Seal up all entry holes, including cracks that extend all the way through floors and walls or that lead to spaces behind

A house cricket can be distinguished by its pale brown color and the three dark bands on its head. It breeds outdoors as well as indoors.

Both house and field crickets live and breed outside, but house crickets can also breed indoors. They lay their eggs behind baseboards, in cracks and crevices, and in refuse. Field crickets entering the house in summer die indoors by fall or early winter.

Outdoors, crickets lay their eggs individually in soil, often in dumps. Field cricket nymphs take about 12 weeks to reach maturity; house cricket nymphs take about 30 weeks.

The field cricket produces as many as three generations each year in warmer states. In cold-winter areas, eggs generally overwinter, although half-grown nymphs may also do so. The house cricket has only one generation per year outdoors but may breed throughout the year indoors, becoming a persistent pest.

Field crickets migrate indoors when their natural food supply of grasses dries up, when cold weather begins in fall, or when rainfall is excessive. During some years, cricket populations explode, and huge numbers may invade homes and other buildings.

Crickets become active at night and hide during the day in dark, warm places. Lights attract them, and at times crickets may literally cover streets and sidewalks under street lamps.

Prevention and Control
You can control cricket infestations indoors by spraying with Dursban® (chlorpyrifos—Ortho Home Pest Insect

Control), Baygon® (propoxur—Ortho Ant, Roach & Spider Killer), or synthetic pyrethrins (Ortho Home & Garden Insect Killer Formula II). Pay special attention to areas along baseboards, in closets, under stairways, and similar places. Don't concentrate solely on the basement and ground-floor level; crickets can fly and may enter second-story windows as well.

When crickets are wandering in from outdoors during summer and fall, an outdoor treatment may be needed. Apply a band of insecticide spray or granules 1 to 2 yards wide around the foundation. Use diazinon (Ortho Diazinon Granules), malathion (Ortho Malathion 50 Insect Spray), Dursban® (chlorpyrifos—Ortho Outdoor Ant, Flea & Cricket Spray or Ortho Home Pest Insect Control), or Baygon® (propoxur—Ortho Pest-B-Gon® Roach Bait, or Ortho Earwig, Roach & Sowbug Bait). Pay particular attention to door thresholds, windows, weep holes in brick facings, and other possible entry points. Also spray under garbage cans and other potential hiding places.

If crickets are a persistent problem, seal up possible entry points, especially near the ground level. (See pages 7 to 11 for more details on preventing crawling pests from entering your home.)

Because lights attract crickets, turn on outdoor lights as little as possible, or use yellow light bulbs, especially during cricket migrations in late summer.

EARWIGS

You may see earwigs indoors hiding around baseboards, under rugs and cushions, and in clothing. They don't damage clothing or other household items, but they may eat cat and dog food and occasionally get into other stored foods. At times they occur in large numbers and can be a serious nuisance.

All earwigs have a somewhat disagreeable odor, but in one species, the striped earwig, the odor is particularly pronounced.

The name "earwig" derives from an ancient European superstition that these insects will enter a sleeping person's ears. Fortunately, this doesn't happen. Earwigs are practically harmless to humans, although if you handle some of them carelessly they can pinch hard enough with their formidable-looking pincers to draw blood. They use these pincers to capture prey and to defend themselves.

The Pest
Over 20 species of earwig live in the United States. The European earwig, first seen in Portland in 1909 and now found in all 50 states, is the most common, except in some southern states where

the three, average slightly less than 1 inch; European earwigs average about ⅝ inch long; and ringlegged earwigs, the smallest of the three, average ½ to ⅝ inch. Ringlegged earwigs have lighter bands around their legs, although the bands may be indistinct.

Earwigs generally travel about by crawling, but some species have wings that are folded up under a protective cover on their backs and occasionally use them to fly short distances at night. Earwigs cannot breed indoors, except in potted plants, and they eventually die inside a house.

Nighttime lights may attract large numbers of earwigs, since they come out at night to feed and hide during the day in moist, protected places under leaves, in cracks in the foundation and soil, in hoses, and in tree wounds, among other places.

Most earwigs eat living and decaying plant material and other insects, although some, such as the striped earwig, feed primarily on insects and only occasionally eat garden plants.

Prevention and Control
Sweep or vacuum up earwigs that have entered your home. Spraying insects directly with

Although they find much more to eat outside—such as insects and dead and living plants—earwigs can infest homes and feast on stored foods.

the striped earwig (also called the shore earwig) and the ringlegged earwig tend to be more common.

The easiest way to tell the difference between adults of these earwigs is by their size. Striped earwigs, the largest of

an insecticide such as synthetic pyrethrins (Ortho Household Insect Killer Formula II) before sweeping makes it easier to collect the insects. Try not to crush them because this increases the unpleasant odor they emit.

Spraying along baseboards and other areas where earwigs might hide or enter controls most earwig problems. Use an insecticide containing Dursban® (chlorpyrifos—Ortho Home Pest Insect Control) or Baygon® (propoxur—Ortho Ant, Roach & Spider Killer). The best time to apply an insecticide is in the late afternoon or early evening so that it is fresh when the earwigs become active.

If earwigs continue to come indoors, treat the outside of your home in a band 1 to 2 yards wide around the foundation, paying particular attention to hiding places under debris and places where they are entering your home. Use a spray or granule containing Dursban® (chlorpyrifos—Ortho-Klor® Indoor & Outdoor Insect Killer or Ortho Outdoor Ant, Flea & Cricket Spray), diazinon (Ortho Diazinon Granules), or Sevin® (carbaryl). You can also use a bait containing Baygon® (propoxur—Ortho Earwig, Roach & Sowbug Bait or Ortho Pest-B-Gon® Roach Bait) or Sevin® (carbaryl). For severe infestations, a much wider general garden treatment with one of these insecticides is necessary.

Eliminating any loose boards, leaf mulches, and other debris along the foundation will aid control and should be done before you apply a chemical treatment. Because earwigs may hide in such items as cut flowers, vegetables, newspapers, and clothes, check these items and shake them out before you bring them in from the garden.

Exclude earwigs and other crawling insects as much as possible by using tight-fitting screening and sealing any holes they may be entering. (See pages 7 to 11 for more details on keeping crawling pests out of your home.)

ELM LEAF BEETLES

The elm leaf beetle is an annoying shade tree pest that may migrate indoors in large numbers seeking a place to hibernate. This serious pest of elm trees may move from nearby trees into your home or garage in the late summer or fall to spend the winter. The insects congregate behind curtains, beneath carpets, in attics, and in other protected places. During warm spells in winter they become active and crawl about the house. In spring, they wake up for good and search for a way to get back outdoors.

These beetles create a nuisance and embarrassment indoors, but they don't damage household items or food, and they don't bite; nor do they stain household goods, as boxelder bugs and clover mites do.

The Pest

Elm leaf beetles measure about ¼ inch long and are yellowish to olive green. A single black stripe runs along the edge of each wing cover.

Beetles hibernating outdoors search for a dry place, such as a woodpile or cracks in bark, to spend the winter. In spring, they fly back to elm or zelkova trees at about the time that the first leaves are expanding. There they chew holes in the leaves, mate, and lay clusters of lemon yellow eggs on the leaf undersides.

The eggs hatch into larvae, which are black at first and later become pale yellow with two black stripes down their backs. These feed on leaves, reaching a length of nearly ½ inch before forming bright orange-yellow pupae at the base of tree trunks or in bark crevices.

Elm leaf beetles produce two to five generations each year, depending on the climate. They can be a serious pest of elm trees, skeletonizing leaves and weakening

Elm leaf beetles are annoying but harmless winter sojourners; they may enter homes in autumn in order to hibernate indoors.

trees. Infested trees lose vigor and may then succumb to an elm disease, such as Dutch elm disease or phloem necrosis. (For more details about elm leaf beetles infesting trees, see Ortho's *Controlling Lawn & Garden Insects.*)

Prevention and Control

Vacuum or sweep up the beetles indoors, and discard the vacuum bag. To kill beetles in hiding, spray along windows, baseboards, and in similar areas with pyrethrins.

In severe infestations, you should control the beetles outdoors to prevent more from wandering in. Spray along the foundation, in any areas where you see beetles accumulating, and along windowsills and doorsills where beetles may be entering. Use synthetic pyrethrins.

Inspect elm trees in early to late summer for signs of infestation by elm leaf beetles. Extensive areas of unsightly chewed leaves and clusters of beetles mean that elm leaf beetles are weakening your valuable trees and may become a nuisance indoors. To prevent the adult beetles from harming the trees and at the same time to prevent populations from building up and later becoming a household problem, spray the trees with Sevin® (carbaryl—Ortho Liquid Sevin®) or Orthene® (acephate—Orthene® Systemic Insect Control). A thorough application, especially to

the undersides of the leaves, will be needed.

In subsequent years, one spray in the spring after the larvae of the first generation have begun feeding (usually about four to six weeks after the leaves have come out) normally provides protection for the entire season. Repeat the treatment later in the summer if the tree becomes infested again.

If the tree is more than 50 feet tall, you may need to hire a pest control operator with power spray equipment to obtain adequate spray coverage of the entire tree.

Seal any openings through which the beetles may enter the home. (See pages 7 to 11 for details on eliminating entry places.) Open attic windows in the spring to allow the beetles to fly out.

FLIES

Many species of fly become household pests. Some of these may seem indistinguishable to you, although they are actually different flies with different habits and thus require different control measures. You may not recognize other types as flies at all, because they resemble large mosquitoes or small moths.

Many people mistakenly think that small flies grow into large flies. All flies are full grown when they emerge from their pupae. Small flies are different pests than their larger, noisier cousins.

The Pest

Crane flies resemble huge mosquitoes and often frighten people. Their size makes them look scary, but they don't bite or cause any harm. In fact, the adult fly you see doesn't eat anything at all. Their bodies are 1 inch long or more, and they have a wingspan of up to 3 inches and very long, delicate legs that come off easily when pulled.

House flies lap liquefied food with their long tongues. As they lap they can introduce disease organisms.

Crane fly larvae, also called leatherjackets because of their tough skins, live in moist soil outdoors, where they feed on decaying plant material. The larvae of a few species feed on grass roots and blades. Some species produce only one generation per year; others may produce several.

Cluster flies enter homes in late summer to hibernate for the winter, gathering together in clusters in secluded areas, especially in wall voids and attics. During warm days from fall to spring, many become active and find their way into the living areas of your home, where they crawl sluggishly on walls or fly noisily around lights and bump into windows, particularly on the south side of your home. They usually die within a few hours from exposure to heat at the window and from lack of moisture.

These flies, slightly larger than house flies, sometimes occur in huge numbers and can be very annoying. Cluster fly larvae feed on earthworms in the soil, so trying to control the flies by eliminating the breeding source is impractical.

Another fly, the face fly, also hibernates indoors, sometimes together with cluster flies. Face fly larvae breed in fresh, undisturbed cow and horse manure no more than a few hours old. They tend to be more of a problem in rural and suburban areas where animals are kept.

You might mistake tiny drain flies, also called moth flies, for small moths because hairs cover their wings and bodies and because they hold their wings rooflike over their backs, as moths do. These flies are often seen mysteriously appearing from sink and bathtub drains. The larvae live in the organic debris that builds up in drains just above the water-filled drain trap, despite the hot water and soap that may travel down the drain. They also live in similar moist places in sewage treatment sites and dirty garbage containers. Saucers with standing water underneath houseplant containers may also be breeding sites for drain flies.

Outdoors, drain flies may breed in birdbaths, shallow pools, and similar sites that have standing, stagnant water. They are so small (just over $1/16$ inch long) that they can penetrate ordinary window screening.

Fungus gnats are small ($1/8$ inch long), delicate, tan to black flies that you may see running across the soil of your houseplants or flying near the pots. They commonly breed in houseplant soil, and most remain near the breeding site, although you may find them throughout your home. They also breed outdoors in damp, decaying compost, including leaf mold and grass clippings. When numerous, the larvae may damage seedling roots or roots on cuttings but don't do great harm to mature plants. The larvae feed on fine roots and root hairs.

Fungus gnats tend to be more of a problem in wet soils with a high proportion of organic matter (compost or manure). They multiply very quickly, completing an entire generation in as little as two weeks under ideal conditions.

House flies, by far the most common indoor fly pest, excite people to action not only because their buzz annoys, but because they easily transmit disease organisms. These dark gray flies are marked with four longitudinal lines on the thorax and are about $1/4$ inch long.

Adult flies often lay eggs in filthy places where disease organisms abound, such as

Vinegar flies, often called fruit flies, are so small they can pass through ordinary screening. They feed on fruit and other fresh foods.

animal excrement, garbage, rotting vegetables and fruits, unsanitary compost heaps, or any warm, moist, decaying organic material. After visiting these sites, the flies may land on your food and transmit the disease organisms.

Flies regurgitate digestive juices directly onto their food and then suck it up in liquid form. They also defecate while feeding, further spreading contamination. If you eat food where flies have swarmed, especially if it has been sitting unrefrigerated, enough disease organisms may be present to make you sick.

House fly eggs hatch after 8 to 10 hours into larvae called maggots, which form pupae three days to eight weeks later, depending on the temperature. The pupal stage can take as few as four days, but in cold climates house flies may spend the entire winter as larvae or pupae.

House flies can travel more than 10 miles from the place they emerge, but most of them stay within a mile or two, going no farther than necessary to procure food or find suitable places to lay eggs. Adult house flies generally live from 15 to 30 days, depending on the temperature. Without food they live only two to three days.

The reproductive potential of house flies is infamous. A female lays from 350 to more than 1,000 eggs during its lifetime. Under the warmest conditions, a fly can complete its life cycle in less than seven days. One scientist estimated that under perfect conditions, if all flies were to live and there was an unlimited food supply, no predators, and continual warm temperatures, one pair of flies could mate in April and by August their offspring and subsequent generations would cover the earth 47 feet deep. Another scientist believes this figure to be a gross exaggeration, however. He concluded that the flies would cover the earth only to a depth of $2^{1}/_{2}$ feet.

If you see small ($1/8$-inch-long) flies hovering over wine, beer, vinegar, or ripe fruit, you are looking at vinegar flies. Some people call these fruit flies, but true fruit flies, such as the Mediterranean fruit fly and similar flies, are larger and not common pests indoors.

Vinegar flies fly to and lay their eggs on ripe or rotting fruit and vegetables, in garbage cans containing even small portions of these materials, and other places where these items may be moist and fermenting or rotting. The female vinegar fly lays about 500 eggs, and because these flies require only 8 to 10 days in warm weather to complete a generation, populations can grow very quickly.

Vinegar flies can be annoying indoors in late summer and fall, since they swarm and fly about kitchens if anything is there to attract them. These pests are so tiny that they can fly right through most screens.

Many other flies come inside homes. The metallic-colored greenbottle and bluebottle flies, also called blowflies, breed in pet droppings, garbage, and dead, decaying animals, including carcasses of animals that may be trapped in wall voids or attics.

Tachinid flies are medium-sized to large flies that parasitize other insects. If they parasitize a caterpillar or some other insect that has crawled into your home or garage, tachinid flies emerge indoors and, like so many other pests, fly to windows trying to get outside.

Flesh flies look like overgrown house flies. The more common ones have a checkerboard pattern on their abdomens. The larvae feed on decaying organic material, such as pet droppings, snails killed by snail poison, and dead animals, or as parasites on other insects.

Some flies, such as deerflies, horseflies, and blackflies, bite people and can be nasty indeed. (For details about these pests, see page 81.)

Prevention and Control

You can easily control isolated house flies and similar flies with a flyswatter, if you have good aim. If flies occur in greater numbers, you may want to spray the room with an aerosol containing d-trans allethrin (Ortho Professional Strength Flying & Crawling Insect Killer), tetramethrin (Ortho Home & Garden Insect Killer Formula II), or pyrethrins (Ortho Hi-Power Indoor Insect Fogger) or resmethrin. These insecticides effectively kill any flies present when you spray.

Resinous pest strips containing DDVP (dichlorvos) are of limited use for controlling flying insects in your home. You can safely hang them in living areas, but not where food is stored, processed, or eaten or where infants or the elderly are confined. To be effective, pesticide strips must be placed in enclosed rooms where the vapors can build

Despite their alarming resemblance to giant mosquitoes, crane flies do not bite or sting.

up; in well-ventilated rooms or on porches, they prove useless. These pest strips effectively control flies in garbage cans. Simply attach one underneath the garbage can lid. Fly control lasts for one to three months.

You can reduce the number of flies coming indoors by applying a residual insecticide outdoors around your home. Spray surfaces where flies usually land or congregate, such as screens and around light fixtures and garbage cans, as well as places where they may enter your home, such as along the edges of screens and doors. Use an insecticide containing Dursban® (chlorpyrifos—Ortho Home Pest Insect Control), diazinon (Ortho Diazinon Insect Spray), or malathion (Ortho Malathion 50 Insect Spray).

You can scatter fly baits containing trichlorfon (Dipterex®) outdoors around garbage cans and near doorways, according to label directions. Outdoor fly traps, either those containing odors attractive to flies or ones using ultraviolet light, reduce fly numbers in a large area but may actually increase the fly population nearby because many flies are attracted to traps but escape being caught.

You can exclude many types of flies from your home by being sure that windows and doors have tight-fitting screens. These simple structures provide effective year-round relief from all sorts of flying insects.

There are several ways to discourage house and bluebottle flies from breeding around your home. Flies swarm and breed around garbage; be sure to remove your garbage at least weekly, because most flies can't complete their life cycle within a week. Always use tight-fitting lids on your garbage cans to help keep flies out. Clean out any food residue left on the sides or bottoms of the cans with boiling water so that flies can't breed there. Lining the cans with newspaper or plastic trash bags also helps keep them clean.

Compost and piles of rotting grass clippings may also attract flies. Be sure that you are turning compost regularly and that it is heating to a high internal temperature, which will kill fly eggs and larvae. If it is not hot enough, cover it with a thick layer of soil. Twice a week, remove pet droppings from the lawn; these can attract huge

Tiny fungus gnats breed in houseplant soil—especially if it is overwatered.

numbers of flies and are a very common breeding site for many fly species. Bury the droppings, or store them in covered containers until final disposal in a plastic trash bag.

Finding and sealing the openings through which cluster and face flies enter your home is the best method for dealing with these flies. Pay particular attention to the southern side of your home's exterior, because flies tend to gather there in the fall. Observing this wall late on a sunny fall day may help you determine where the flies enter your home.

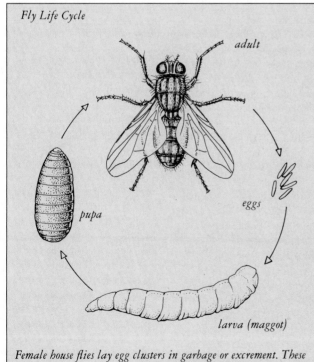

Fly Life Cycle

adult

pupa

eggs

larva (maggot)

Female house flies lay egg clusters in garbage or excrement. These hatch into larvae, which feed, then migrate to a cooler, dryer spot to pupate. The entire life cycle can take as little as 10 days, or the adults may not emerge until the pupae have overwintered.

Fill spots where you see flies entering and any other cracks along the south wall, especially around window and door frames. Tighten any loose shingles. Screen any openings, such as attic vents and air conditioner intake registers. Seal any cracks inside your home that may be letting cluster flies enter through the wall voids.

Spray cracks and crevices and other areas outside where the cluster flies may be hiding. Control crane fly larvae in moist soil and decaying plant matter with diazinon (Ortho Diazinon Soil & Turf Insect Control).

Drain flies don't fly very far, so if you see swarms of them indoors or out, look closely to discover nearby spots where the larvae are living. Control the larvae in drains by removing the organic buildup. To do so, either flush the drain with sink-cleaning materials followed by hot water or thoroughly brush the drain with a long-handled brush.

Control fungus gnats in houseplant soil by drenching or very thoroughly spraying

Large numbers of cluster flies may enter your house in the fall. They typically die en masse trying to escape through a closed window.

the soil with an insecticide containing diazinon. You can often avoid fungus gnat problems by not overwatering houseplants.

Vinegar flies are small enough to pass through ordinary window screens (usually 16 mesh), so excluding them from your home proves rather difficult unless you install fine-mesh screens (24 mesh). An easier method is finding and eliminating whatever either attracts adults or provides breeding places for the larvae. Check for spoiling fruit or vegetables, includ-

ing sacks of potatoes and onions. Just one rotten fruit or vegetable in the bottom of a bag can support a sizable population.

If you leave fruit such as peaches and pears to ripen at room temperature, you can trap vinegar flies by leaving ½ inch of wine in the bottom of a long-necked wine bottle. The flies follow the scent into the bottle, but can't find their way out. Set the trap near the ripening fruit and you'll reduce the numbers of these bothersome pests.

FLOUR MOTHS AND BEETLES

Certain moths and beetles live prosperous and contented lives in stored food, eventually making it unfit for consumption. These pests infest a surprisingly wide variety of food, including flour, spices, cereal, dried beans, dried fruits, nuts, candy, and even pet food. Some of these insects eat almost any kind of stored food; others have a limited diet.

An infestation of flour moths or beetles often gets started when you unknowingly bring home infested food. The food may contain only the tiny, nearly invisible eggs of the insects; even if larvae or adults are present, however, you are not likely to notice just a few, especially if they are hidden in the middle of the package. Some of these insects also feed on nonfood materials, such as wool and leather, later dispersing to stored food. Kitchen pests also fly or crawl in from many different outside sources.

The Pests

The most common moths found in stored food are the Indian meal moth, the Mediterranean flour moth, and the Angoumois grain moth. The Indian meal moth has a ⅜-inch wingspan. The forewings are coppery to purple on the tips and have a broad gray

band near the body. The Mediterranean flour moth has a wingspan of ¼ to ½ inch. Its forewings are pale gray with wavy black lines; the hind wings are cream-colored. The Angoumois grain moth is the largest, measuring ½ to ¹¹⁄₁₆ inch across. Its wings are buff to gray and fringed with long hairs; the hind wings taper to a fingerlike projection. You may see any of these moths flying in your home in the evening, particularly in poorly lit areas and in the soft light of a television. They live only one to two weeks and do not feed but mate and lay eggs in or near stored food.

The white or cream larvae are about ½ inch long when fully grown. They may live a

Adult Indian meal moths have a ⅝-inch wingspread. The larvae infest stored food, spinning silk tunnels as they eat their way through.

month or more before pupating, depending on the temperature. Food infested with moth larvae eventually becomes interlaced with the silken threads they leave as they crawl about. These pests often crawl far from the infested materials when they are ready to pupate. A complete life cycle takes as little as six weeks if the weather is warm and there is ample food.

The most common beetles in stored food include the red flour beetle, the confused flour beetle, the sawtoothed grain beetle, the cigarette beetle, and the drugstore beetle. They are all are about ⅛ inch long and brown, reddish brown, or black. Some of these beetles live for as long as a year or more, laying a few eggs every day.

Flies Spread Disease

Scientists have found in house flies more than 100 different organisms that cause disease in humans or animals. Disease organisms found in flies include those causing diarrhea, dysentery, typhoid, cholera, intestinal worms, polio, eye infections, yaws, anthrax, and tularemia. However, no one knows how often flies actually transmit these organisms from one place to another and how often this transmission actually results in an infection. Unlike malaria-causing organisms, which are transmitted only by certain mosquitoes, many of the diseases that are carried by flies can also be transmitted by other methods, such as contaminated water, unwashed hands, and blowing dust.

Flies certainly present more of a health hazard where conditions are unsanitary and disease organisms abound than they do in clean environments. Although they may transmit food-spoiling organisms, the food must also be left out in warm temperatures before the bacteria can multiply to harmful levels.

Health professionals know that outbreaks of diarrheal diseases correspond to seasonal increases in fly numbers. Experiments in several rural towns have shown that controlling flies reduces sickness from dysentery. The number of cases of dysentery in pesticide-treated towns decreased, while the number of cases in similar towns left unsprayed remained the same. The incidence of dysentery reversed itself when, after 18 months, the sprayed towns were no longer sprayed and the unsprayed towns were sprayed.

Unlike pantry moths, both the adult and the larval stages of pantry beetles eat stored food. The light cream to yellowish white larvae measure from ⅛ to ¼ inch when fully grown. They don't spin any silken threads, and they pupate in the same food they eat. A complete life cycle can take as little as five weeks for some species, but other species require a minimum of several months.

Bean and pea weevils develop only within the dried seeds of legumes such as beans and peas. They do not attack other food products.

Prevention and Control

The first step in controlling pests in stored food is to find all the places where they are breeding. Search the cupboards and all other food-storage areas thoroughly. Telltale signs of an infested

Confused flour beetles live mainly on flour and grain products. Check newly purchased food for infestations before you store it in your kitchen.

bag or box include tiny holes or webbing in food or packages. Pay close attention to opened packages and boxes of food that have been on the shelf for a long time. Any package that gives the slightest indication of being infested should be held suspect.

If you find only one infested package, the infestation could be relatively new. Removing that package may be all you need to do. Most often, however, by the time you notice insects they have already spread to other food packages.

Mediterranean flour moths are common in flour, but also infest such items as nuts, beans, and chocolate. Their webs sometimes clog flour mill machinery.

Place any packages that are obviously heavily infested in a plastic bag, and throw them in the garbage. You need not throw out packages that are only lightly infested or only suspected of being infested; eat these within a few weeks, but keep them in the refrigerator to prevent any insects from maturing.

You can kill insects in stored food by freezing or heating it. Keeping the food in the freezer for three days kills most insects. Or spread the food out on a cookie sheet in the oven at 120° to 130° F for two hours. Be careful not to scorch the product. You can probably obtain the desired temperature in a gas oven merely by turning up the pilot light. In electric ovens leave the door open slightly to keep the temperature from rising too high. Use the freezer technique for nuts, as heat may cause them to become rancid.

To kill insects infesting dried fruits, drop the fruit in boiling water for about one minute. Spread the fruit to dry before re-storing it.

In a serious infestation, not all insects will be in contaminated packages. Adult moths may already have left the package in search of other places to lay their eggs. Larvae of certain insects commonly leave infested food and search for a crack or crevice in which to pupate. Pests also may be breeding in food spilled on or behind shelves.

To control these insects outside the packages, thoroughly clean the cupboards. Remove the shelves, if possible, and look for pests on the edges of the shelves and the shelf supports. Vacuum or sweep the cupboards and shelves, and discard the insects and crumbs in a sealed plastic bag. Then scrub the shelves with soap and hot water. Spray insecticide along the edges on both the tops and bottoms of the shelves, in all floor and wall corners, and in any other cracks and crevices. Use tetramethrin and sumithrin (Ortho Home & Garden Insect Killer Formula II) or pyrethrins. These are both short-lived insecticides that can be used safely in

This dog biscuit has become infested with larvae of the drugstore beetle—an insect said by one entomologist to eat "anything except cast iron."

food cabinets if you remove the food and utensils first.

In a serious infestation, you might consider applying an insecticide that has a longer residual effect, such as Dursban® (chlorpyrifos—Ortho Roach, Ant & Spider

Killer) or diazinon. Apply it to cracks and crevices only, and allow the spray to dry before replacing food or utensils.

Ridding your pantry of pests will take continuous, persistent effort. Because some pests can live for many weeks without food, the threat of reinfestation exists until all of them die off or are killed. It is best, especially for several months after eliminating the infested products, to store any susceptible food in containers with tight-fitting lids. Buy only small quantities of the type of foods found infested so that you will consume them quickly. Long-term storage always encourages problems. Finally, keep storage units dry. Moisture favors the development of some pests, and dryness discourages them.

MILLIPEDES

Millipedes live outdoors and occasionally wander into homes, sometimes in large numbers. They usually die inside within a few days because of the dry conditions and lack of suitable food. Millipedes feed primarily on rotting leaves and wood and other kinds of moist, decaying plant material. They do not damage anything indoors, nor do they bite. If they are picked up, a few species will exude an irritating liquid.

At certain times millipedes migrate in huge numbers and may swarm into basements and other rooms near the ground floor. These migrations most often occur in the fall when the millipedes are looking for a suitable place to spend the winter. They also migrate en masse when their food supply has been depleted or their habitat has become too dry or too wet; during this type of migration millipedes often travel in an uphill direction.

The Pest

Millipedes are red-brown to black and cylindrical or slightly flattened. Some can be 4 inches long at maturity, but most are less than an inch. The distinctive feature of these "thousand-legged worms" is their numerous tiny legs—two pairs on each body segment. Centipedes (see page 79) are often confused with millipedes but have only one pair of legs on each body segment. Centipedes are predators that quickly run from danger; millipedes move slowly and when prodded are more likely simply to coil up and remain still.

Millipedes are not insects but are a close relative. They need to live in a moist environment because their bodies lack an outer wax

Millipedes move slowly and curl up into this defensive position when disturbed.

layer, which keeps insects from dehydrating. They also have no way to close their breathing tubes, and so they lose moisture easily.

Females lay batches of eggs just below the soil surface. The young millipedes look like the adults except that they have no more than seven segments and only three pairs of legs. They molt many times as they grow, adding segments and legs. Many species of millipede reach sexual maturity in two years, but some take up to five years and then can live for several more years.

Millipedes are most abundant on the shaded side of a house, where it is darker and

more moist. They are most active at night and hide during the day in the soil or beneath stones, boards, mulch, or plant debris.

Prevention and Control

Millipedes move slowly, and indoors they can be easily swept or picked up with a vacuum cleaner. If they continue to migrate inside, spray areas where they may enter, such as around doorsills and windowsills in the basement or on the ground floor. Also treat cracks and crevices, particularly in dark areas. Use Baygon® (propoxur—Ortho Hornet & Wasp Killer) or pyrethrins.

Ideally, millipedes should be controlled outdoors before they enter. Treat moist areas around the foundation where millipedes accumulate and near basement doors and windows where they may enter. In severe infestations, treat a 5- to 20-foot band around the entire foundation and in other areas of the yard where they may be breeding. You may need to re-treat this band if great numbers of millipedes pass over it, because they use up the available insecticide. Use diazinon (Ortho Diazinon Insect Spray), Dursban® (chlorpyrifos—Ortho Outdoor Ant, Flea & Cricket Spray or Ortho-Klor® Indoor & Outdoor Insect Killer), or a bait containing Baygon® (propoxur—Ortho Pest-B-Gon® Roach Bait).

To keep populations down, remove or dry out moist places, such as compost piles and accumulations of leaves or other plant debris, especially if they are near the foundation. Take measures to reduce excessive moisture in crawl spaces and basements (see page 57 for details). Exclude the pests where possible by sealing up potential entry points (see pages 7 to 11 for details).

PSOCIDS

Psocids (pronounced *so*-kids) are also called booklice because you sometimes see them around mildewed books and papers in damp buildings. The name "booklice" is misleading because these pests do not damage books or other items, and unlike lice they cannot bite humans or animals.

Psocids feed primarily on microscopic molds growing in damp, warm places. These molds, and the accompanying psocids, can become particularly abundant on upholstered furniture stuffed with straw or other natural vegetable fibers and on certain starchy adhesives used in bookbinding and wallpaper. To a lesser extent, psocids scavenge dead insects and insect eggs. Other insects, typically silverfish or cockroaches, damage books. In most cases psocids are an annoyance simply because they occur in large numbers.

Psocids can infest food stored in slightly damp locations. They feed on molds growing on these foods and to some extent on the food itself. They contaminate the food with their bodies but do not transmit disease. Psocids commonly infest starchy items and dry goods.

New homes and other buildings sometimes are host to large populations of psocids for several months after construction. Because moisture, especially from plaster and green lumber, takes time to dissipate, psocids find such structures to be ideal living spots. They may first multiply in the wall void or places where moisture builds up and mold growth flourishes. Afterward, they may enter the living space.

The Pest

These almost colorless to light gray or brown pinhead-sized pests may catch your eye because of their characteristic jerky manner of walking.

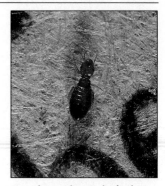

Psocids may live in books, but probably won't harm them. They eat mainly molds growing on damp surfaces in homes.

Those found indoors are wingless, but many psocids that occur outdoors under bark or leaves have wings that they hold rooflike over their bodies. Immature psocids look like miniature adults.

Psocids lay eggs in cracks. The nymphs hatch about one week later, under ideal conditions, and reach maturity in about a month. Adults live about three months. Most indoor psocids continue to breed throughout the year, but a few species have only one generation per year.

Psocids multiply indoors in the spring and summer but virtually disappear during the winter when the indoor humidity drops. Preferring dark, undisturbed places, psocids seek warmth (75° to 85° F) and high humidity (75 to 90 percent). Eggs can survive a cold, unheated room during winter, but the adults die under these conditions.

Prevention and Control

Because both psocids and the molds they feed on require high humidity to survive, reducing moisture by using a dehumidifier, repairing leaks, and similar measures helps control both problems. A relative humidity of less than 50 percent for several weeks or more will control most psocid infestations. A lower relative humidity (for example, below 30 percent) is even more effective. Artificial heat or a long period of hot, dry weather also controls psocids.

Spray with pyrethrins along baseboards and other places where you see psocids to provide immediate control of exposed insects. Since they may be in the wall void, respray if you see more.

Control infestations in furniture by drying it thoroughly in sunlight for several days or by replacing the stuffing material with a synthetic substance. You can also have the furniture fumigated by a pest control operator or spray it thoroughly with an insecticide containing pyrethrins.

Discard infested food, or spread it out in a thin layer on a cookie sheet and place it in an oven at 130° F for two hours to kill the psocids. Stir occasionally to ensure rapid heat penetration. Drying out the cupboard or pantry helps control psocids. Store susceptible foods in airtight containers. If the infestation is widespread in the kitchen, remove or cover all food and utensils and spray or fog the infested area with pyrethrins (Ortho Hi-Power Indoor Insect Fogger).

SILVERFISH

If you find irregular holes chewed in papers, books, or clothes, suspect silverfish. These pests are particularly fond of starchy material, such as the glue used in bookbindings and to hang wallpaper, the sizing mixed in paper, or the surface glaze used to give paper a glossy look. Silverfish also feed on cotton, linen, and synthetic fibers, especially if any of these fabrics are starched. They seldom injure wool, hair, or fibers of animal origin. You may see a yellow stain, tiny scales, or excrement on paper or fabric on which silverfish have fed. These pests also feed on a wide variety of stored food, especially rolled oats, wheat flour, breakfast cereals, and dried beef.

You may first notice silverfish when you lift an object under which they were

Shiny silverfish scurry into crevices when a light is turned on. These silverfish are of the species known as firebrats.

hiding and see them dart for cover. They occasionally appear in bathtubs and sinks because they fall in and are unable to climb the slippery sides to escape. Hiding in dark, tight places during the day, silverfish come out only at night to search for food and moisture.

Silverfish get into homes and buildings by being carried in on cardboard, books, papers, or food that has come from infested sites. Once inside, they move to other rooms by crawling along pipes and through cracks in walls and floors.

Many other pests, especially cockroaches, crickets, clothes moths, and carpet beetles, also damage paper and/or clothes. Control measures vary according to the pest, so be sure you have correctly identified the culprit.

The Pest
There are two main species of silverfish, each of which has slightly different preferences as to temperature and humidity. The name "silverfish" refers to the entire group of insects, but the most common species is also known simply as the silverfish. This species prefers damp, cool places (75 to 95 percent humidity and 70° to 80° F) and is common in basements and around water pipes. The other type of silverfish, called the firebrat, prefers damp, warmer places (70 to 80 percent humidity and 90° to 102° F). You are more likely to see firebrats around ovens, heating units, and hot-water pipes.

The silverfish is a shiny silver or grayish insect that is less than 1/2 inch long at maturity. The firebrat is mottled gray, whereas the silverfish is uniformly gray. Both species have scaly, tapered bodies with two slender antennae in the front and three long, thin appendages in back, which make them look like fish.

The females lay eggs throughout most of their adult life but lay fewer eggs if disturbed. They deposit their eggs in cracks and under boxes and other objects. The

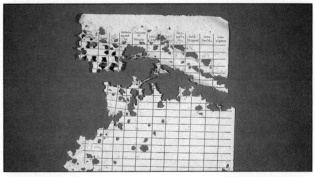
Silverfish often feed on paper, preferring sheets with starch sizing or a glazed surface.

nymphs are smaller versions of the adults and reach sexual maturity in as little as two to three months under ideal conditions, although they may take as long as two years. Some can survive for up to a year without food.

Prevention and Control
Control silverfish with a spray containing diazinon or Dursban® (chlorpyrifos—Ortho Home Pest Insect Control). Treat areas such as cracks along baseboards, door and window frames, closets,

behind drawers and shelves, behind and around the edges of bookshelves, and where pipes go through walls. Dusts such as boric acid or silica aerogel also provide effective silverfish control.

Silverfish bait packets are available in some stores, and you can use them instead of sprays. However, they take longer to control an infestation. They do not attract silverfish from very far away (perhaps 1/2 inch), so to be effective they must be spaced as close as every 1 to 3 feet.

Treating only the warmer parts of the building is usually sufficient to control firebrats.

If you still find silverfish or firebrats two or three weeks after a treatment, they are probably coming from areas you missed. You need to make a second, more thorough treatment.

To prevent silverfish, throw out unnecessary books, magazines, and papers because they may be important breeding places. Move books around in bookcases periodically. Clean up crumbs, and store food in tightly sealed containers. Plastic bags that zip shut are effective for storing food or valued papers or clothes. Eliminate sources of excess moisture by repairing leaky plumbing and by using proper ventilation, especially in rooms that tend to be damp, such as basements, laundry rooms, and bathrooms. If practical, reduce the number of areas where silverfish hide and breed by sealing up cracks and crevices around pipes and baseboards.

SOWBUGS AND PILLBUGS

Sowbugs and pillbugs sometimes come indoors, where they gather near sources of moisture, such as planters, damp crawl spaces, leaky pipes, or damp basements or bathrooms. They may annoy you, but they don't harm household items, nor do they bite or spread disease. They feed primarily on moist, decaying plant and animal matter. Sowbugs and pillbugs

Pillbugs roll up when disturbed and lack the pointy tails of sowbugs, their close relatives.

can't survive more than a day or two indoors unless they find moisture.

Unlike insects, sowbugs and pillbugs breathe through gills and lack a protective layer of wax to prevent them from losing water. To conserve moisture they usually remain under damp compost piles, leaves, boards, and other objects that keep them from drying out. At night they take advantage of the lower temperatures and higher humidity to travel in the open.

The Pests
Sowbugs and pillbugs are more closely related to crayfish, shrimp, and crabs than they are to insects. They have seven pairs of legs and are covered with a series of armorlike gray plates. When they are full grown, they may reach 1/2 inch long.

The difference between sowbugs and pillbugs is that pillbugs can roll up into tight balls to protect themselves;

sowbugs can roll up only partially and therefore are more likely to go scurrying for cover when disturbed. They both inhabit gardens throughout the United States.

Female sowbugs and pillbugs deposit their eggs directly into pouchlike structures called marsupia on the undersides of their bodies. The young are white when they hatch and continue to live in the pouch for six to eight weeks after the eggs were laid. Depending on the temperature, moisture, and food availability, sowbugs and pillbugs produce one to three generations a year and can live more than two years.

Prevention and Control
You can sweep up or vacuum sowbugs and pillbugs indoors, since they don't move very fast. Or spray them with Dursban® (chlorpyrifos—Ortho Home Pest Insect Control), Baygon® (propoxur—Ortho Ant, Roach & Spider Killer), tetramethrin and sumithrin (Ortho Household Insect Killer Formula II), or d-trans allethrin (Ortho Professional Strength Flying & Crawling Insect Killer).

To prevent them from coming indoors, apply a band of insecticide around the foundation and to any damp surrounding areas. Use a bait containing Sevin® (carbaryl) or Baygon® (propoxur—Ortho Pest-B-Gon® Roach Bait) or a spray or granule containing diazinon (Ortho Diazinon Insect Spray or Ortho Diazinon Granules) or Dursban® (chlorpyrifos—Ortho-Klor® Indoor & Outdoor Insect Killer or Ortho Outdoor Ant, Flea & Cricket Spray). If you know exactly where the sowbugs enter your home, such as cracks underneath doors or windows, apply the insecticide directly to those places. You can also apply a bait or granule directly to the areas in the yard where you see them.

Eliminating the moist environment that enables sowbugs to survive, both indoors and outdoors next to the foundation, may be all that is necessary to control the situation. Reduce any excess watering around the foundation, especially of ivy and other ground covers. Remove organic matter, such as accumulations of leaves, compost piles, old boards, and so forth where sowbugs and pillbugs breed. Repair leaky pipes and other sources of moisture indoors. Damp crawl spaces will benefit from better ventilation (see page 57 for details).

You can exclude these creatures from your home by sealing or blocking possible entry points such as doorsills, below-grade windows, and cracks in foundations or brick facings. Weather stripping or caulking may be sufficient.

SPIDERS

Spiders are one of the most common pests in and around homes, yet many people fear them more than any other household pest. Fortunately, this fear is greatly out of proportion to the danger they present. Spiders may look frightening, but they rarely even try to bite people, and the vast majority couldn't penetrate skin if they tried to bite. Spiders with bites that can puncture skin have such weak venom that the bite is usually far less painful than a mild bee sting. The black widow and the brown recluse are two species with stronger venom. (See page 88 for more information on these spiders and on spider bites.)

Spiders produce thin strands of silk, which they use in a variety of ways. Many use the silk to build webs to trap their prey. The webs may be very neat and symmetrical, with a spiral of strands supported on spokes that radiate out from the center; the webbing may form a funnel; or it

Spiderwebs are unsightly in homes but catch intruders, such as these hornets. The light-brown objects suspended in this web are spider egg cases.

may simply be a tangled mass of strands with no particular form or shape.

Some species of spider hunt their prey and don't build webs to trap them. These hunting spiders, such as wolf spiders, jumping spiders, crab spiders, tarantulas, and trapdoor spiders, wander about searching for insects or lie in ambush waiting for them to pass by. Hunting spiders, particularly males, occasionally venture indoors searching for prey or mates.

Spiders do many things with silk besides building webs. They use it to make egg sacs, trapdoors, and protective retreats. Many spiders use a line of silk as a dragline to drop down to a lower level. You might walk into one of these lines long after the spider has used it. Young spiders disperse by letting out a silken strand that catches the wind and balloons them up into the sky. They have been sighted as high as 5,000 feet and hundreds of miles out over the ocean. The masses of ballooning threads often seen on fall days are called gossamer.

Spiders feed entirely on other spiders and insects. They tend to be more numerous in homes that have a plentiful supply of other pests that they can eat. A spider quickly immobilizes its prey by injecting venom into it or wrapping silk around it. If a spider has eaten recently, it will save its catch for a later meal. Before eating an insect, the spider injects it with a powerful digestive saliva that turns much of the body tissue into liquid. After feeding, it drops the empty carcass to the ground. Mature spiders can survive for many months without eating.

Most of these pests aren't dangerous, but nevertheless most people would prefer to keep them out of the house. Spiderwebs, droppings, and cast carcasses can create messy corners in rooms, garages, and basements.

Daddy longlegs, or harvestmen, are not true spiders, but are close relatives. They cannot produce silken strands, and their diet consists mainly of small living or dead insects. Daddy longlegs do not bite people.

The Pest

Spiders are not insects but arachnids. You can easily distinguish spiders from insects because spiders have eight legs rather than six, two main body parts (a head and abdomen) instead of three, no antennae, and no wings. Their body shape resembles that of mites, ticks, and scorpions more than insects.

Each female spider lays as many as several hundred eggs inside a silken egg sac and then stays to guard its eggs. A female may make several egg sacs during its life. The young hatch within three weeks during warm weather. They are cannibalistic, so they stay together for only a few days before dispersing. Depending on the species and climate, spiders either mature into adults the same year or overwinter and mature the next season.

Prevention and Control

Knock down spiders, together with their webs and egg sacs, and crush them with your foot, or spray them. Use Baygon® (propoxur—Ortho Ant, Roach & Spider Killer) or Dursban® (chlorpyrifos—

Ortho Roach, Ant & Spider Killer or Ortho Home Pest Insect Control). When spraying for spiders that stay on their webs, you should try to hit the spider with the insecticide directly. Spraying the walls or floors next to them is not as effective because spiders usually remain on their webs and therefore won't have an opportunity to walk over the sprayed surface.

Keep in mind that dusty webs have long since been abandoned by the spider. Spraying these webs will not control spiders. Regular cleaning of your home helps prevent spider populations from building up and removes their webs.

If there are many spiders and their prey indoors, an aerosol that automatically releases all of its contents in an empty room, such as Ortho Hi-Power Indoor Insect Fogger, may be useful.

Controlling spiders indoors may be all that is necessary, because many indoor spiders don't live outdoors. However, if spiders or their prey are coming in from outdoors, spray around the outside of the foundation, under eaves, and at other entry points, such as around doors and windows. You can also use this treatment to control spiders that breed outdoors around the home but do not come indoors. For outdoor spider control, use an insecticide that contains Dursban® (chlorpyrifos—Ortho-Klor® Indoor & Outdoor Insect Killer) or diazinon (Ortho Diazinon Insect Spray).

Seal up any cracks and crevices through which spiders can enter your home. (See pages 7 to 11 for details.) Eliminate any places next to the foundation where spiders would be able to hide, such as accumulations of trash and plant litter. In severe cases, consider trimming shrubs and trees so that they are well away from the house.

SPRINGTAILS

These pests, found in moist indoor places, spring several inches into the air when disturbed. They don't damage anything indoors or transmit diseases. However, some people working in areas where there are large numbers of springtails develop a mild itching if the pests crawl or hop on their skin.

Springtails feed primarily on algae, fungi, and decaying plant material, such as leaves and logs. A few species can damage young, tender plant parts in close contact with the ground. They can multiply indoors in houseplants.

Infestations of springtails indoors are confined to moist places, such as damp basements and bathrooms, around drains or water pipes, and along windowsills where there is condensation. You may also find them in houseplant soil or in the saucer underneath a plant. They can become abundant around swimming pools and drown in large numbers, requiring frequent cleaning of the pool.

Springtails sometimes invade houses searching for moisture, especially if their moist breeding places outdoors begin to dry up. They enter through and around screens, under doors, and through vents. Outdoor lights sometimes attract them and increase the likelihood that they will find their way inside. They die soon after entering the home unless they find sufficient moisture; however, additional insects may continue to enter.

The Pest

Springtails are so small (about 1/16 inch long) that you seldom notice these whitish, grayish, or bluish insects.

Springtails get their ability to jump from a special appendage that they fold underneath their bodies. It is held in place with a catch mechanism, and when released it strikes against whatever the insect is

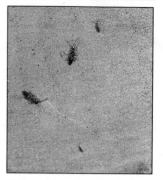

Harmless, but annoying in large numbers, springtails may inhabit damp areas of a home.

resting on, hurling it several inches into the air.

Little is known about the life cycle of springtails. They lay eggs in damp leaf mold and similar locations, apparently including damp places indoors such as dirty floor drains. They complete several generations a year.

Prevention and Control

The most effective way to control springtails is to eliminate moisture. Simply airing out the area with a fan often works. Fix any water leaks or other sources of excess moisture. You can use a vacuum cleaner to collect springtails from floors or windowsills.

If springtails persist indoors, or if it is not practical to dry out the areas they inhabit, spray infested areas with an insecticide containing diazinon.

You can eliminate springtails in houseplant soil by allowing the soil to dry out. If they persist, take the plant outside and treat the soil with diazinon (Ortho Diazinon Granules).

If springtails are coming from outdoors, control them by drying out the soil around the foundation or by treating places along the foundation where they enter with diazinon (Ortho Diazinon Soil & Turf Insect Control). You can eliminate springtails around swimming pools by allowing the soil in nearby landscaped areas to dry out or by treating the soil in these areas with diazinon (Ortho Soil & Turf Insect Control).

STRUCTURE AND WOOD PESTS

An invasion of wood-destroying pests in your home is a serious matter. If these destructive pests go unnoticed or are left unchecked, controlling them later and repairing the damage they have done can be expensive. This is especially true if the pests cause extensive damage to structural parts of your home. Although they usually damage wood slowly and it often takes a number of years to cause extensive trouble, more often than not, wood pests get the time they need. They can go undetected for years. Regular inspections of your home can turn up signs of these wood destroyers early enough to prevent serious damage.

Some of these pests, such as termites and wood-boring beetles, actually feed on wood. Others, such as carpenter ants and carpenter bees, use wood only for nesting. Whatever their reason for invading wood, these pests often have fairly specific preferences. How long ago the wood was cut, the amount of moisture it contains, the kind of tree it came from, and whether it came from the inner heartwood or the outer sapwood all may influence whether a given pest will attack.

Although the vast majority of wood-damaging pests are insects, some particularly important types of noninsect wood pests—decay and wood-rot fungi—are also discussed in this chapter. These fungi not only damage wood and leave it structurally unsound, they greatly increase the chance that other wood pests will invade. Wood rots weaken wood, making it more attractive to insects such as carpenter ants and subterranean and dampwood termites, which prefer soft wood. The fungi

Carpenter ant workers keep the young ants sorted, placing eggs and the different stages of larvae and pupae in different chambers.

Carpenter bee

Less visible than other household pests, wood-destroying pests can literally eat you out of house and home before you notice them. Learn the telltale signs of these pests and how to prevent costly damage.

also provide additional nitrogen and vitamins to nourish the insects. Scientists have proven that the odor of certain decay fungi attracts subterranean termites.

Wood pests enter your house in a number of ways. Subterranean termites may be colonizing a tree stump or pieces of wood left beneath a newly built home. From there they have easy access to the timbers of the house. Insects such as powderpost beetles and drywood termites may come indoors uninvited on antique furniture, wooden crates, or other wood

items. The pests may even be living inside paneling or cabinets when they are installed in your home. Wood pests also crawl or fly indoors through cracks and crevices in the foundation or around windows and doors; once inside, they can begin new colonies.

This chapter tells you how to identify most wood pests, even if all you see is the damage they are causing. More important, it explains how you can help prevent termites and other pests from attacking. You will find out how to decide if it is practical for you to attempt to control these pests yourself and find out what control methods to use.

Before you try to control any of these wood pests, be sure you have properly identified the culprits. The effective control for one kind of termite, for instance, may not be successful for another kind. If, after reading about these pests, you are still uncertain of the pest's identification, seek additional help. You can take or mail a piece of the damaged wood and one or more of the insects, if you can get them, to an entomologist at your county or state cooperative extension service. A reputable pest control company will also be able to identify a pest, as will a private consulting entomologist. The pest control operator will provide you with an estimate for control and repair.

Although you can control many of these wood pests yourself, in some cases the eradication procedures are best left to a professional pest control company. To get rid of certain termites and beetles the entire house or individual pieces of furniture may need to be fumigated—procedures which legally can be carried out only by a licensed operator. Professional pest control operators can also inject insecticides into concrete slabs or foundation walls to get at termites.

BEETLES

Wood-boring beetles are second only to termites in the amount of damage they do to homes and wooden items. Their larvae can cause severe damage to furniture and structural wood as they tunnel and eat their way through it, growing larger as they go.

These beetles can migrate to your home from outdoors. They often breed in dead tree limbs and stumps, prunings, firewood, and similar items. Nearby houses can also be sources of infestation, as can new and antique furniture, wooden crates, and similar wooden items, especially if they have sat in a warehouse or storage facility and have unpainted or unvarnished surfaces. New cabinets and paneling can be infested with powderpost beetles, especially if they were unknowingly made from infested wood.

If you discover wood-boring beetles in or around your home, you will need to identify the pest properly

Adult old house borers emerge through oval exit holes, leaving a powdery mess inside timbers.

before you can know its potential for damage. Generation after generation of certain kinds of beetle can reinfest a single board, eventually reducing it to little more than a mass of holes and sawdust. Others limit their attack to fresh wood to which the bark is still attached. These beetles usually die as soon as the wood dries out, but their tunnels remain in the wood permanently and may cause a great deal of unnecessary concern.

Sawdust pours from exit holes made by old house borers in this crawl space. The larvae may feed for two to ten years before pupating into adult beetles.

The Pest

Powderpost beetles include lyctids, anobiids, and bostrichids. Their common name reflects their behavior: Most species infest and reinfest seasoned wood, eventually reducing it to powder. The old house borer is also sometimes called a powderpost beetle because it too can reinfest wood.

Lyctid (pronounced *lick*-tid) powderpost beetles feed in the sapwood of seasoned hardwoods, such as ash, oak, walnut, and hickory. Common targets are hardwood flooring, paneling, furniture, gun stocks, tool handles, and firewood. Lyctids prefer recently dried wood rather than wood that has been finished for 10 years or more.

Wood infested with lyctids has round holes $\frac{1}{32}$ inch to just over $\frac{1}{16}$ inch in diameter on its surface. The holes are loosely packed with an extremely fine wood powder that feels as smooth as talcum powder. There are no pellets or larger chunks of material. Piles of powder may accumulate below the exit holes.

Lyctid beetles lay their eggs in the fine pores on the surface of wood. They never lay eggs on waxed, varnished, or painted wood, because the pores are sealed. The larvae hatch from the eggs and bore into the wood. They are so small at first that you can't see the entrance holes they make. They create wandering tunnels in the interior of the

wood, finally forming pupae just below the surface. The emerging adult beetles bore out of the wood, leaving the characteristic exit holes. A complete life cycle takes one to two years.

Adult lyctids are $\frac{1}{8}$ to $\frac{1}{4}$ inch long and reddish brown to black. They are most active at night, and this, along with their small size, is the principal reason they are so seldom seen. Because light attracts them, you may find dead beetles on windowsills. Even during a heavy infestation indoors, however, you may never see an adult.

The whitish larvae reach $\frac{1}{4}$ inch long at maturity. Because the larvae remain inside wood and blend in well with sawdust, you are not likely to see them unless you very carefully break apart and inspect a piece of actively infested wood. Lyctid adults and larvae are very similar to anobiids and bostrichids, so you'll probably need an entomologist to make a positive identification.

Anobiid (pronounced an-o-*bee*-id) powderpost beetles can attack almost any kind of softwood or hardwood, but they feed most often in lumber or items made from the sapwood of softwoods. These pests commonly damage structural wood, such as joists and studs, in crawl spaces and basements. Anobiid beetles become problems more often in moist basements rather than very dry ones. You only occasionally find them in the structural wood of drier areas

such as attics. Some anobiids also damage furniture and other wooden items.

Infested wood has round holes $\frac{1}{16}$ to $\frac{1}{8}$ inch in diameter on its surface, made by the adult beetles as they exit. The holes are loosely packed with a gritty powder that is much coarser than the powder made by lyctid beetles. Some of this powder may accumulate below the exit holes.

Anobiids occasionally lay their eggs on surfaces that have been painted or varnished, but they normally choose unfinished surfaces. They may also lay an egg in the mouth of an old exit hole. Like lyctid larvae, anobiid larvae create wandering tunnels in the interior of the wood and then pupate just below

Exit holes are often the only sign of powderpost beetles. Larvae (right) are hidden and adults (left) are most active at night.

the surface. Adult beetles bore out of the wood, leaving the exit holes, and lay eggs to renew the cycle in the same or a different piece of wood.

Anobiid adults and larvae resemble lyctids in color and size. You are unlikely to see them. A complete life cycle takes one to four years. They develop most quickly in warm, moist areas.

A few species of anobiid beetle make tapping sounds in the early evening. The sound comes from the adult hitting its head against wood as a mating call. Depending on the species and how quiet the house is, you may be able to hear this sound and trace it to a piece of infested furniture. One particularly loud species called the deathwatch beetle is so named because superstitious people used to believe the clicking sound was a foreboding of death. The sounds are most common in the late spring when the adults emerge and mate.

Bostrichid (pronounced, bos-*trick*-id) powderpost beetles do not cause problems as often as lyctids and anobiids do. Most attack hardwoods, including bamboo, but some species also feed in softwoods. They rarely reinfest seasoned wood. Bostrichids seldom cause significant damage to structural wood; they primarily infest individual pieces of hardwood flooring, trim, or furniture.

Most bostrichids are about the size of lyctids and anobiids and leave round exit holes ranging from ⅛ to ⅜ inch in diameter. However, some of these beetles are much larger. The black polycaon beetle, for instance, ranges from ½ to ¾ inch in length and leaves exit holes well over ¼ inch in diameter.

A sawdustlike powder that ranges from fine to coarse, depending on the species, fills the tunnels of bostrichid beetles. The powder tightly packs the tunnels and does not sift out of the wood easily. Unlike lyctids and anobiids, adult bostrichids bore into the wood and lay their eggs directly in the tunnels. A complete life cycle usually takes about a year.

Ambrosia beetles, which infest only live wood, should not be confused with powderpost beetles. These pests are easily identified by their characteristic exit holes, which are lined with a dark blue or black stain caused by ambrosia fungi. Some of the stain diffuses into the wood surrounding the holes. The holes are also all a uniform size, between ¹⁄₅₀ and ⅛ inch in diameter, and are clean, without any frass or sawdust.

Roundheaded and flatheaded borers, with the exception of the old house borer described later, infest only living or dying trees outdoors—they can be serious plant pests. Likewise, bark beetles tunnel just beneath the bark in living trees and in log cabins in which the bark has been left on the logs. You may see damage caused by these insects in lumber or boards milled from trees that are infested, but by that time

Above: Larvae of wood-boring beetles weaken furniture and structural wood by eating tunnels through it.
Below: Firewood can be infested with wood-boring beetles. These logs show exit holes made by powderpost beetles.

they can cause no further damage. Usually, the beetles have left long before the wood is in use. Wood infested with round- or flatheaded borers has round or oval holes from ⅛ to ¼ inch in diameter in its surface. Usually, some of the tunnels are exposed lengthwise in lumber because they were made before the wood was milled. The tunnels are often tightly packed with fine or coarse sawdustlike material.

The old house borer is the only roundheaded borer that infests and reinfests wooden items made of seasoned softwood. It especially favors pine and occurs as far west as Minnesota and Texas. Contrary to its name, the old house borer is more common in homes that are less than 10 years old, rather than in older homes. It may get into structural wood anywhere in a house. As with the other powderpost beetles that infest and reinfest wood, if this pest is not controlled, it can eventually reduce wood to a powdery mass encased in a thin

shell of wood. Additional damage results when the adults bore out of paneling, flooring, plasterboard, and other materials.

Infested wood eventually shows oval exit holes ¼ inch in diameter in its surface. However, the first noticeable sign of old house borers may be the sound of the older larvae chewing in the wood. They make a rhythmic ticking or rasping sound, much like a mouse gnawing. In severe infestations the sawdustlike frass, which they pack loosely in their tunnels, may cause the thin surface layer of wood to bulge out, giving it a blistered look. The frass is a mixture of very fine powder and larger, blunt-ended chunks. You may find small piles of it beneath infested wood.

The females lay their eggs in cracks and crevices on the surface of the wood. The larvae tunnel in the wood for 2 to 10 years before pupating, so you may not see evidence of their damage for that long. They develop most quickly under warm, humid conditions in new or fairly new

Is the Infestation Active?

Before you go to the trouble and expense of controlling wood-boring beetles, you need to determine whether an infestation is still active. Because you seldom see the insects themselves, you must rely on other evidence. First, check the sawdust. If it is yellow and partially caked, the infestation is probably no longer active. If any of the sawdust is fluffy and light-colored, like freshly sawed wood, the infestation is still active. Powder streaming from the exit holes is also a sign of recent activity.

If you are still unsure whether you have an active infestation, place a dark paper or cloth under the wood and leave it for at least several weeks from May to September. Check for new sawdust piles on the paper. A little sawdust may sift out of the current holes because of natural vibrations and should not be confused with new activity.

Wood-boring beetles always tunnel inside wood; the only visible holes you see appear when the adults bore straight out of the wood. If you do see winding tunnels exposed at the surface of the wood, it's a sure sign that the tunnels were made before the wood was cut. This gives you a useful bit of information that helps you establish when some of the tunnels were made. The beetles that made those particular tunnels may have died long before the wood was ever placed in service.

The round holes in this hardwood panel were excavated by powderpost beetles.

wood. Adult beetles emerge from the wood in June and July, mate, and lay eggs to begin a new cycle.

Adult old house borers are brownish black and ⅝ to 1 inch long. A characteristic wavy, light-colored line crosses the wing covers from side to side. The larvae are cream colored and up to 1⅛ inches long at maturity.

Prevention and Control
You can sometimes intercept wood-boring beetles before they do any damage to your home by carefully inspecting old and new furniture, lumber, firewood, and other wooden items before you bring them indoors or store them next to your home. Look for bulges, exit holes, and sawdust.

Because lyctids and, to some extent, anobiids do not lay eggs on waxed, painted, or varnished surfaces, you can prevent an infestation from spreading by finishing all wood of the same type as that the beetles are infesting. If larvae already infest the wood you treat, expect to see some exit holes until they have all matured and left. Fill the holes with putty, and apply the finish over the holes.

Wood-boring beetles will not attack pressure-treated lumber. Kiln drying kills whatever beetles are infesting the wood at the time, although they may reinfest it.

Anobiid infestations are not as prevalent in dry crawl spaces and basements as they are in damp ones. Maintain good ventilation in the crawl space, and do everything else possible to keep the area as dry as possible. Do not place fine-mesh screening over vents in crawl spaces, because this decreases air flow through the area—¼-inch heavy screen will deter animals and allow adequate air flow. (See page 12 for information on controlling moisture.)

A common means of controlling extensive infestations of wood-boring beetles in buildings is to tent and fumigate the entire structure. Because fumigation requires special skills and equipment, only licensed pest control operators can do the job. They will use either methyl bromide or Vikane® (sulfuryl fluoride). These insecticides can also be used to fumigate a single piece of furniture if you suspect that the infestation is limited to that one item. You can treat a limited infestation in a woodpile or in lumber outdoors by spraying or painting the infested boards with Dursban® (chlorpyrifos—Ortho-Klor® Soil Insect & Termite Killer) according to label directions. Wait at least two weeks before burning the wood. This treatment will not kill insects deep inside the wood, but will kill them as they bore out of it, preventing them from laying additional eggs on the wood. Wood thoroughly treated with this product will be protected from further attack for a number of months.

If you choose this kind of treatment but the infestation is widespread or is in a tight, enclosed area, such as a crawl space, it is best to have a professional pest control operator apply the insecticide. These professionals have been trained to make this type of application.

The advantage of fumigation over directly treating infested boards with an insecticide is that fumigation controls the beetles everywhere in the structure, even in areas that you might overlook or that are hidden from view. The disadvantages are that fumigation leaves no residual protection after the tent has been removed, and if only a small area is infested, fumigation is more expensive than applying a surface treatment.

If you discover a localized infestation early in its development, removing the object or infested boards may be all that is required. If you choose this method, carefully inspect all other wood in your home to be sure that the infestation has not already spread, and reinspect the wood twice a year to see if pests reappear.

CARPENTER ANTS

Although most ants can be extremely annoying pests, they do little actual damage. Carpenter ants are another story. They can pose a serious problem because of their potential to harm the structure of your home. A large colony nesting in the house may eventually cause considerable destruction. If you find and eliminate carpenter ants at an early stage, you may be able to avoid costly repairs.

Carpenter ants do not eat wood for food; they simply excavate it to create a nesting place. They commonly colonize wood in wall voids, floors, structural timbers, and ceilings. Like other ants, carpenter ants can also be a nuisance when they move about in the house searching for food. They feed on a wide variety of sweets in the kitchen, as well as meats and grease. Outdoors, they eat small dead and living insects, but their primary food is honeydew, the sweet substance excreted by aphids and related insects that suck juices from plants.

Finding a few carpenter ants in the house does not necessarily mean that they are nesting in your home. However, you probably do have a nest somewhere indoors if you see large numbers of carpenter ants on a continual basis or in the spring. You can be certain that they are nesting indoors if you also see piles of fibrous sawdust, often containing pieces of ant bodies, spilling out of cracks or through slitlike openings they make in the wood.

Another sign of infestation is large numbers of winged ants swarming indoors, although these could also be other types of ants. On warm days in the spring and early summer, carpenter ants, like other ants and termites, send out numerous winged reproductive ants from their nests. If they've emerged indoors, these ants frequently collect at windows.

Some carpenter ant colonies produce a faint rustling noise that is loud enough to be heard in a quiet room. The sound may increase when you vigorously rap the wood where they are nesting. Placing your ear or a stethoscope against the wood can sometimes give you an idea of exactly where they are nesting. Because carpenter ants are most active from about 8 p.m. to 4 a.m., and especially around midnight, the sound is most audible then.

Because of their small size and their tendency to choose already-hollow places in which to nest rather than excavating additional areas, it sometimes takes many years before carpenter ants cause significant damage.

Carpenter ant workers vary greatly in size. They excavate large galleries in wood to make nesting sites, but do not actually eat the wood they remove.

The Pest

These black or reddish black ants measure from ¼ inch to more than ½ inch long and are among the largest ants to infest homes. Their jaws are also large. The life cycle of carpenter ants is similar to that of other ants (see page 32 for details).

Carpenter ant galleries resemble the work of termites, but they have an almost polished, sandpapered appearance and lack debris. The galleries generally run with the grain of the wood. The ants show a strong preference for moist, softer wood that has begun to decay, but some species attack newly built structures. They will also nest in wood surrounding existing hollow places, such as wall voids, spaces under outdoor siding, and various cracks and crevices.

These ants are especially likely to attack houses located in wooded areas and forests. The initial colony may be formed indoors by a founding queen, but more frequently a mature colony or a section of a mature colony migrates into the man-made structure from a nest built in a declining tree or stump outdoors. These satellite colonies are slower to cause severe damage.

Carpenter ants often cross over to a home on tree or shrub branches that are rubbing against its roof or side. They then enter any crack or hole they can find. Sometimes entire colonies are unknowingly brought indoors in firewood, but these rarely establish themselves.

Carpenter ant colonies grow slowly. They reach maximum size in three to six years, at which time they begin producing their first winged reproductives. A mature colony contains 2,000 to 3,000 or more ants, a small number compared to the typical colony size of many other types of ants and subterranean termites.

Prevention and Control

Ant baits do not control carpenter ant colonies effectively. Sprays applied to their trails kill some ants and stop them from using the trail but do not kill the queen or eradicate the colony.

To succeed in controlling carpenter ants, you must find the nest or nests and treat them directly. This requires a very careful and thorough inspection, because there may be more than one nest, and the nests tend to be difficult to locate. For an extensive or serious infestation, consider hiring a licensed pest control operator to do the inspection and control work.

Locate the nests indoors by looking for piles of sawdust and listening carefully for the faint rustling sound a colony makes at night. If you find winged ants indoors in the spring, try to locate the hole or crack where they emerge. Because these ants tend to excavate moist wood, carefully inspect areas of your home where there may be excess moisture, such as around pipes and bathroom plumbing, under a leaky roof in the

attic, around fireplaces, and in the basement or crawl space. Pay particular attention to any wood that is in direct contact with soil.

Once you have found the nests, treat them by drilling into the wood or void area and applying an insecticide directly into the nest. Drill and treat about every foot in and around the suspected area to intercept all the galleries. Also drill into and treat any void areas next to the infested wood, and treat cracks and crevices near the nest.

You can use either a dust or a spray, but a dust is preferable because it drifts farther into the galleries. Use Dursban® (chlorpyrifos) dust (Ortho-Klor® Indoor & Outdoor Insect Killer) or liquid (Ortho-Klor® Soil Insect & Termite Killer), diazinon, Sevin® (carbaryl), or Baygon® (propoxur).

You should also inspect outdoor areas within 100 feet of your home. Examine dead trees and tree stumps, living trees that have scars or other dead areas, bushes, woodpiles,

and similar places for carpenter ant nests. Control these nests by using any of the insecticides listed for controlling ants outdoors (see page 32). Treat the nest directly as well as the area surrounding the nest, the trunks of any infested plants, and, especially, the entire foundation of the house.

An integral part of carpenter ant control is to correct all moisture problems in your home to make wood less desirable to the ants. Refer to the section on decay rots on pages 56 and 57 for more information. Also trim trees and shrubs so that no branches touch or come close to any part of your home.

To control carpenter ants and prevent them from entering your home, it is a good idea to remove dead trees, stumps, and branches within 50 feet of your house. Store firewood up off the ground and well away from your home. Burn any insect-infested firewood immediately, or discard it.

Insect Problems in Firewood

A host of insects can live in firewood. The cut wood not only provides food for most of the wood-eating pests described in this chapter, it also acts as a cozy, dry place for many overwintering insects and spiders. When you bring firewood into your warm home on a cold winter day, many of these insects will start moving around because of the warmth. Only in rare cases—such as when bringing in an entire colony of termites or carpenter ants, or introducing a pest such as powderpost beetles that can reinfest unfinished wood surfaces—will any insects from firewood become serious household pests. Most likely, firewood insects will simply be a nuisance as they crawl around.

To avoid an infestation by insects brought in in this way, it is best to burn up any firewood you bring indoors within 24 hours. Although it may be convenient to store

more firewood indoors, the risk of infesting your home with carpenter ants, termites, or borers, though small, isn't worth it.

Locate woodpiles at least 10 feet from your house. Storing the wood against the side of your home gives insects easier access and increases the chance of their infesting the structure. Try to keep the pile as dry as possible with sturdy plastic sheeting or another covering. Space out logs, bricks, or cement blocks lengthwise under the pile to keep the firewood off the ground.

If firewood is heavily infested with insects, you can treat the woodpile with insecticide. Spray with Dursban® (chlorpyrifos—Ortho-Klor® Soil Insect & Termite Killer) or diazinon (Ortho Diazinon Insect Spray). Wait at least two weeks before burning wood you have treated, or follow label instructions.

CARPENTER BEES

Carpenter bees burrow into exposed, dry wood in buildings, outdoor furniture, and fences. Several holes do not seriously affect the structural integrity of the wood, but carpenter bees tend to be attracted to the same site each year. Over a period of several years, they may cause extensive cosmetic damage. These bees prefer painted or unpainted softwoods, such as redwood, cypress, cedar, and Douglas fir.

The Pest

Carpenter bees are robust, yellow-and-black insects about ½ to 1 inch long. They are very similar to bumblebees, for which they are easily mistaken. Carpenter bees, however, nest in wood and lead solitary lives; bumblebees usually nest in the ground in colonies.

To distinguish a carpenter bee from a bumblebee, note the pest's abdomen. On a carpenter bee, the top side of

Developing from successively laid eggs, each of these carpenter bee larva resides in a separate, sealed cavity in the tunnel.

the abdomen is shiny and mostly bald, whereas a bumblebee's abdomen is dull on top and completely covered with short hairs.

Male carpenter bees have the alarming habit of buzzing around the heads of people who come near their nesting sites. Since they lack stingers, they are entirely harmless. The females do have stingers but rarely use them.

A telltale sawdust pile marks the spot where a carpenter bee bored into this porch railing; flaking paint exposed the wood and made it more inviting.

Carpenter bees bore holes ½ inch in diameter. Freshly drilled holes have a rather coarse sawdustlike substance below them. The tunnel usually goes about an inch deep and then abruptly turns at a right angle and travels in the same direction as the wood grain. Most tunnels extend 4 to 6 inches, but there are records of tunnels as long as 10 feet, the result of several years of reinfestation.

The carpenter bee spends the winter in an old nest tunnel and emerges in the spring. After mating, the female searches for a nesting site. It may use an old tunnel without extending it, lengthen an old tunnel, or bore an entirely new one. The female packs a mass of pollen and regurgitated nectar (a mixture called bee bread) at the end of the tunnel and lays a single egg on it. It then seals off this portion of the gallery with a disk of chewed wood pulp. The bee repeats this procedure until it has made about six sealed cells in a row. The larvae feed on the provisions, form pupae in their cells, and finally emerge as adult bees from six weeks to three months later.

Between late summer, when most carpenter bees emerge, and fall, the young adults drink nectar and provision abandoned tunnels with pollen. They feed on the pollen from time to time when the weather is cold,

eventually settling down in the tunnels to hibernate for the winter. An occasional bee may enlarge its hibernating quarters, but they don't bore any new holes until spring.

Prevention and Control

Control carpenter bees by spraying insecticide directly into the entrance holes. Use Dursban® (chlorpyrifos—Ortho-Klor® Soil Insect & Termite Killer) or Sevin® (carbaryl). Do not plug these holes immediately, but give any adult bees a chance to pass freely in and out, so that they spread the insecticide farther into the gallery. Several days after treating, plug the holes with putty, caulking compound, or dowels. Paint or varnish this and other exposed wood surfaces to prevent carpenter bees from drilling additional holes or returning to treated ones. If you plug the tunnels without first killing the insects, the bees trapped inside will bore new openings.

You can sometimes discourage carpenter bees by painting or varnishing all exposed wood or by using pressure-treated wood for building projects. They often favor stained cedar or clapboards over painted surfaces.

DECAY AND WOOD-ROT FUNGI

Various kinds of fungi grow on and in moist wood; some can cause considerable damage to a structure in a short period of time.

Dry rot is the term most often used for decay fungi attacking houses and other structures, although this term is misleading. At the time the fungus is growing, the wood is noticeably damp because the fungus carries water to the wood through water-conducting tubes. Like all other decay fungi, dry rot fungi cannot grow on dry wood. As soon as you dry the

wood, the fungus stops growing and further damage ceases. The name "dry rot" probably originated because by the time you discover the rot the fast-growing fungi may have died and the rotten wood dried up.

Dry rot can appear almost anywhere in a structure but is most common in exterior trim, in the subflooring where leaks occur around plumbing, in crawl spaces, in decks, and anywhere that wood is exposed to excess moisture.

Decay rots attack all kinds of wood, but certain woods, including western red cedar, bald cypress, and redwood, tend to be more resistant. Sapwood of all trees, even the fairly resistant ones, is always more susceptible to rot than heartwood. Heartwood, the older, often darker wood closer to the center of the trunk, has more rigid cell walls and contains certain chemicals that protect it somewhat from decay.

Wood rotted by decay fungi eventually becomes weak and structurally unsound. Some kinds of decay fungi also produce an odor that can attract termites.

High humidity increases the chance of wood rot, but it alone does not provide enough moisture for decay. The source of moisture must be more substantial, such as a plumbing leak or condensation. Wood in direct contact with moist soil is especially likely to be attacked.

Molds and mildews are other kinds of fungi that grow on wood. They can weaken wood-fiber products (such as chipboard or particle board), but they do not damage other kinds of wood. Mildew does help wood absorb additional moisture, however, and, if it continues to grow, provides a good sign that the wood is moist enough for more-destructive decay fungi to attack. Like decay fungi, molds stop growing as soon as wood dries out.

The Pests

Wood rots are fungi, which are parasitic plants. The fungi consist of threadlike strands called hyphae, which aggregate into masses called mycelia. Under ideal conditions, the hyphae produce fruiting bodies that release reproductive spores. These disperse in the air, and if they come into contact with moist wood give rise to new hyphae.

Wood thoroughly rotted by a decay fungus is characteristically dark brown, severely shrunken, and crumbly. It has crevices that separate the wood into small, cubical pieces as it breaks apart. This is often referred to as checking.

Most decay fungi attack wood that is already moist, but some can carry water to otherwise dry wood. They do this with thin, water-conducting tubes formed by the mycelia. These rootlike, white or black strands may grow together in thick bundles up to 1 inch in diameter. The mycelia make it possible for the fungi to colonize wood away from obvious sources of moisture. All decay fungi need moisture to continue growing, however.

Decay rots grow fastest when the temperature is between 60° and 90° F. Temperatures held above 115° F for a long time kill them, but they can survive freezing.

Prevention and Control

Once wood rot sets in, there is no remedy other than replacing the damaged wood if it is structurally unsound. You can take steps to prevent it from causing further damage, however.

The best way to stop or prevent wood decay is to keep wood dry and the humidity low. Dry wood has the added benefit of being less attractive to certain wood-damaging insects, such as termites and carpenter ants.

Most decay problems indoors develop in the bathroom and the kitchen. Thoroughly inspect all sinks,

Wood-rot fungi have left this timber with characteristic checking—cracks in a square pattern.

shower stalls, and toilets for any signs of moisture from leaking pipes or loose caulking, and repair these moisture sources as soon as possible.

Outdoors, check for plugged or leaky gutters, a common source of moisture. Downspouts should divert rain water away from the foundation, and the ground should be sloped so that water flows away from your home. Check the caulking around windows and chimneys periodically to be sure that water can't seep in. Keep all exterior wood, especially trim, well painted and free of cracks or chipped paint, which allow rain to soak into the wood, thus encouraging rot. Check for wood that is in direct contact with the soil.

Well-insulated, energy-efficient homes may be too airtight, which can encourage wood rot on wood indoors. Where air can't circulate, moisture from showers, houseplants, laundry, and so on builds up the indoor humidity. In winter, the excessive humidity can condense on colder surfaces and bead up on walls, windowsills, and windowpanes. If you use a humidifier to raise indoor humidity in the winter, condensation problems may increase.

Condensation occurs because warm air can hold more water vapor than cool air. When warm air is suddenly cooled, such as along a windowpane or some other cold surface, it can't hold

as much water vapor, and it dumps the excess as free water (condensation droplets). Wood rot resulting from condensation has become an even greater problem during the past two decades because people are making their houses more airtight to be more energy efficient.

It is important to keep humidity low enough indoors to help keep wood from absorbing excess moisture and to decrease condensation problems. Cooking, washing clothes, and showering are major sources of water vapor inside most homes. Be sure that the clothes dryer is properly vented to the outside. When you cook or shower, open a window or use an exhaust fan.

You can greatly reduce the humidity in crawl spaces, where wood rot can be a problem, by having good cross-ventilation. As a general rule, you should have 1 square foot of vent area for every

Leaky pipes can provide enough moisture to encourage wood decay fungi to grow.

150 square feet of floor space. The vents should be located opposite one another, spaced along the walls and near the corners for optimum ventilation. Do not allow garden plants to block the free flow of air through the vents. In humid crawl spaces where good ventilation is impractical, a moisture barrier laid over the soil has proven highly effective in reducing humidity. Many different kinds of material will work, but polyethylene sheeting 4 to 6 millimeters thick and roll roofing are most commonly used. Because wood shrinks

as it loses moisture, if the soil is exceedingly wet you should begin by covering only part of it so the wood will dry slowly and will not warp.

You can reduce the humidity in basements by using an air conditioner or a dehumidifier during the humid months. These appliances will help prevent condensation on cold-water pipes and air conditioner ducts.

Also check to see that your attic vents aren't blocked. Good air circulation from an attic fan or properly spaced and installed gable and soffit vents prevents moisture buildup in attics.

No wood under your home and porch should be in direct contact with soil. Where wood absolutely must rest against soil, or where it will be subject to constant moisture, use timber that has been pressure-treated with a preservative. This treatment causes the preservative to penetrate deep into the wood and makes it more rot resistant than wood that is painted or dipped into a preservative. In wood that is only painted or dipped, cracks or breaks in the surface can give decay fungi access to the untreated wood beneath. If you do paint preservative on, two or even three coats provide better protection.

Cutting through pressure-treated wood exposes an untreated inner core. After cutting pressure-treated wood, apply preservative to any cut surfaces that lack the darker color caused by the pressure treating.

TERMITES

Termites damage more wood than all other wood-infesting insects combined. These social insects often go unnoticed in a home for years while they slowly eat whatever wood they have access to. By the time you or a pest control operator discover them, these pests may have significantly damaged your home. However, if you discover and control termites early in an infestation, you may be able to avoid spending thousands of dollars to get rid of them and repair the damage.

Termites eat anything composed of cellulose (the hard part of plants), including wood, cardboard, and paper. No wood is wholly resistant to them. The heartwood of redwood, southern cypress, and some junipers are initially resistant to most termites but become more susceptible once the repellent chemicals leach out. No redwood is resistant to drywood termites.

The Pest

A termite colony is a highly coordinated system composed of different castes, each of which carries out specific duties. Worker termites are the numerous ones you see when you break into a colony; they are wingless, sexless, and blind or nearly blind. Workers excavate wood and care for the other termites. Soldiers have well-developed heads with large mandibles; they defend the colony against ants and other intruders. Soldiers are much less common than workers. Like workers, they are practically blind, are wingless, and cannot reproduce. The workers must feed the soldiers because they are incapable of feeding themselves.

The reproductive caste is composed of several kinds of termites. The queen and king are the primary reproductive individuals in a colony. The queen is much larger than the other termites and basically does nothing but produce eggs. Some tropical termite queens can produce as many as 3 million eggs per year; a queen can live for about 25 years. (Workers have a life span of only two to three years.) Workers constantly clean and feed the queen. The king termite stays with the queen, and they continue to mate during the king's life.

A mature colony also contains many supplementary reproductives. These do not produce as many eggs as the queen, but if the colony has many of them, they can produce significantly more eggs as a group than the

The undersurface of these wooden stairs appeared relatively intact until exploration with a screwdriver revealed extensive subterranean termite damage.

queen can. A termite colony could theoretically exist forever because the supplementaries take over egg production if the founding queen dies. Supplementaries also may migrate with some workers and soldiers and begin a new colony.

Swarmers are winged male and female termites that emerge from a colony en masse at certain times of the year. They fly a short distance, after which their wings fall off. A male and female then pair off, search for a place to start a new colony, and mate about a week later. As with winged ants, swarmer termites are rarely successful. The overwhelming majority are eaten, never find a mate, or never find a suitable food and nesting site.

A termite colony develops slowly. The queen lays only a few eggs during its first year. After three or four years, the colony begins to produce a few winged swarmers. It takes even longer before the colony produces the supplementary reproductives that greatly boost the growth rate of the colony. Evidence of structural damage may take even longer.

The only time a termite colony can greatly damage your home in only a few years is when an already-strong existing colony invades the structure. A subterranean termite colony, for instance, could develop first in a nearby tree stump; in large, dead roots; in buried wood forms used in construction; or in a similar food source and from there eventually find a way to invade your home.

Termites are often mistaken for ants, and vice versa. Both live in colonies, often in the ground, and produce winged reproductives that swarm at certain times of the year. (See the illustration on the opposite page for help in distinguishing termites from ants.) Ants have elbowed antennae and thin waists, whereas termites have relatively straight antennae and fat waists. The swarmers also have different types of wings. A termite's fore- and hind wings are about the same length and are filled with many fine veins. An ant's forewings are about a third longer than its hind wings, and both wings have only a few large veins.

Before you attempt to control termites, you need to know what kind you have— subterranean, Formosan, dampwood, or drywood. Each one has distinctive habits that require different control procedures.

Subterranean termites are the most widespread termites in the United States. They occur in every state except Alaska and are especially common in the southeastern and Pacific Coast states. They were once rarely found in the northern states, but centrally heated buildings have enabled them to survive where normally the cold, dry weather would have killed them.

A mature colony can be very large, containing as many as 50,000 individuals. Subterranean termites maintain their colonies underground and build characteristic mud tubes, called shelter tubes, across and into any surface between the ground and their wood food source. The tubes protect the termites from such enemies as ants and help conserve moisture while the insects travel from their below-ground colony to their food source above.

Signs of a Termite Infestation

Here are four signs that indicate that termites are infesting your home.

- Swarmers appear indoors, even though the windows and doors are closed. Because swarming may be over within a few hours, you may only notice large numbers of termite wings near the emergence site or on windowsills.
- Wood is so thin you can push a screwdriver into it. Frequently, such wood will bulge or blister before it actually caves in. Underneath the surface you will find termite galleries, but you may not see the termites themselves. Poking suspicious-looking wood

with an ice pick can help you find hollowed wood.
- Mud tubes, usually going from soil to wood and attached to concrete or some other surface for support, are a sign of subterranean termites. Look also for soil used to seal off openings in cracks and crevices.
- Piles of fecal pellets indicate drywood termites. These pellets are uniform in size and shape and at first glance look like tiny seeds. They have six sides and rounded ends and are less than 1/25 inch long. The termites push them out through small "kickout" holes or cracks in the infested wood.

Subterranean termites must maintain contact with the ground to survive, so they quickly repair broken tubes. If the termites in the wood are not able to return to the ground, they usually die, except in rare cases when they have survived on some other continual source of moisture.

You can distinguish a colony of subterranean termites because there is soil in the gallery, the eaten-away portion of the wood. Subterranean termites are also the smallest termites found in North America. The nymphs are white and 1/4 inch long; the swarmers are dark and about the same size or a little more than 3/8 inch long. Swarming usually occurs after the first rainy day in autumn or on a warm, humid spring day.

The Formosan termite is an important pest that was introduced to North America from the Far East after World War II. It has long been a serious pest in Hawaii and has established a foothold in harbor towns such as Charleston, South Carolina; Galveston, Texas; and several other locations in Texas, Louisiana, Mississippi, Florida, and Tennessee. This pest could eventually become established all along the southern, southeastern, and southwestern coasts and in the lower Mississippi Valley.

Formosan termites are a vigorous and aggressive subterranean termite species. They are known to tunnel through asphalt, rubber, lead, mortar, and plaster to get to wood on the other side. They can cause more damage faster than other species because their colonies may contain a few million individuals.

Although Formosan termites are similar to other subterranean termites, they are different in several ways. Formosan swarmers are larger—up to 5/8 inch long— and the soldiers have oval-shaped heads, whereas the heads of other subterranean termites are more rectangular. Unlike other subterranean termites, Formosan termites swarm much later in the day and are strongly attracted to light. They usually begin swarming just before sundown and continue until about midnight. Other

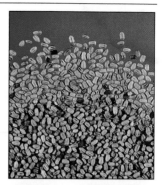

Above: Subterranean termite galleries contain soil. Soldier termites have longer and darker heads than the workers.
Below: Drywood termites do not need contact with soil and their galleries do not contain soil. This chamber is occupied by two reproductive termites.

Above: Drywood termite pellets are sculptured with characteristic ridges. Piles of the pellets appear below infested wood.
Below: Most swarming termites do not succeed in starting colonies; however, large numbers of them in or around a home may warrant a termite inspection.

subterranean termites rarely build nests without tubular soil connections, but Formosan termites can do so if there is a constant water supply.

Formosan termites also produce a hard material called carton. This mixture of saliva and wood resembles a sponge and lines the galleries of these termites. It is sometimes found between infested studs and in the underground portions of their nests.

Dampwood termites occur mostly in the cool, humid areas of northern California, western Oregon, and Washington. They are found to a lesser extent in other regions of the far West.

Homes with leaky plumbing or those situated on moist soil, such as beach houses and forest cabins, are the most prone to attack by dampwood termites. These pests start their colonies in damp, usually decaying wood, and from there they may extend their galleries into sound, relatively dry wood. Unlike subterranean termites, dampwood termites do not require contact with the ground and do not build shelter tubes. Inside their galleries are scattered small fecal pellets similar to those of drywood termites; these measure about 1/25 inch.

Dampwood termites don't have a worker caste. White nymphs, the immature termites, carry out all the work. These eventually mature into either soldiers or reproductives. The nymphs, which are up to 1/2 inch long, are by far the most common individuals in a colony. The yellowish brown swarmers are about 1 inch long, including their wings, making the dampwood termite the largest of the three kinds of termites.

A mature colony of dampwood termites has several thousand individuals. Swarming occurs most often on warm, sultry days in late summer or early fall, just

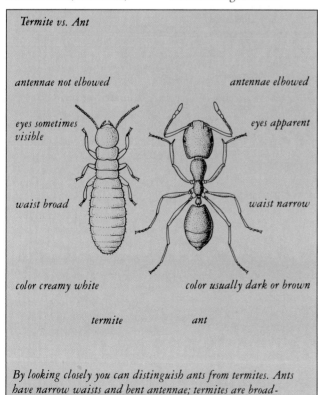

Termite vs. Ant

antennae not elbowed

antennae elbowed

eyes sometimes visible

eyes apparent

waist broad

waist narrow

color creamy white

color usually dark or brown

termite

ant

By looking closely you can distinguish ants from termites. Ants have narrow waists and bent antennae; termites are broad-waisted with relatively straight antennae.

before sunset, and sometimes after the first fall rains. They are strongly attracted to light.

Drywood termites occur mainly along the southern rim of the United States, especially in Arizona, southern California, along the Gulf Coast, in Florida, and along the lower Atlantic seaboard. They also occur in Hawaii. Because drywood termites can survive in furniture, they will occasionally be found in all regions of the country in furniture that has been transported from an area where they occur naturally.

Drywood termites do not need contact with the soil or with any other moisture source. They are even more common in dry roof rafters than in areas near the ground. These pests also quite frequently infest door and window frames. Typical entry points include joints between rafters and sheathing under eaves, openings around window frames, attic and crawl-space vents, and similar places. They also infest trees, entering through dead limbs or other injuries.

Drywood termite galleries are perfectly smooth and clean except for the characteristic hard fecal pellets scattered in the chambers. Unlike subterranean termites, drywood termites don't build mud tubes, and they don't bring soil into their galleries.

Drywood termites are larger than subterranean termites and smaller than dampwood termites. The white nymphs are up to $\frac{3}{8}$ inch long, and the reddish-brown-headed swarmers are up to $\frac{1}{2}$ inch long, including the wings. Swarming usually occurs on sunny days when the temperature is 80° F or higher, usually in the summer or fall. Rain does not stimulate their swarming, as it does with subterranean termites.

Drywood termites, like dampwood termites, have no worker caste. The nymphs eventually become either reproductives or soldiers. A mature colony has between 2,000 and 3,000 individuals.

Prevention and Control

Subterranean and Formosan termites cannot survive in wood without access to moisture (normally soil moisture). Depriving the termites of moisture both prevents them

Subterranean termites create mud tubes made of dirt or wood and a kind of glue. Without contact with soil or protection within the mud tunnels, these termites cannot survive.

from becoming a problem and can control them if they have already invaded the wood. A three-pronged attack is usually necessary to achieve this goal.

First, prevent any direct contact between structural wood and soil. Most subterranean termite infestations start where wood touches the ground. Such wood is moister, making it subject to decay fungi, which also makes the wood more attractive to termites. Termites can also easily enter the wood undetected if it touches the ground. Wood needs to be separated from soil by at least 18 inches. Use pressure-treated lumber for construction wherever wood must touch soil.

Subterranean termite colonies often become established in wood debris that is buried or touching the ground, and then the termites enter your home. Remove all such wood, especially nearby stumps, buried chunks of roots, and buried concrete form boards that were used in construction. If your home has a crawl space, remove from it any pieces of wood, cardboard, or paper large enough to carry or be raked up. Also be sure to remove any paper collars left around pipes and plumbing.

The next step is to eliminate excess moisture. Use the same procedures you use to prevent decay rots (see page 157). Leaking plumbing, too few or blocked crawl-space vents, and poor drainage are some of the common sources of excess moisture that encourage decay rots and termite attack.

Your next step is to create a chemical barrier. The first two steps alone may be all you need to prevent or control an infestation, but applying an insecticide greatly improves your chances of successfully dealing with termites. It also provides you with additional insurance against future problems. When you are dealing with a severe infestation, an insecticide treatment is usually essential.

You may be able to apply the insecticide yourself. If you have a particularly severe infestation or one that is complicated or difficult to treat because of insufficient room in your crawl space, however, contact a licensed termite operator to do the job for you. A professional has the skills and equipment to do these kinds of difficult jobs effectively and safely.

A termite operator usually creates a chemical barrier by injecting an insecticide into the soil on both sides of the foundation and around any supporting piers and pipes. The insecticide must be placed deep into the soil to stop the termites from entering any hidden cracks that develop in the foundation or slab. Often, the exterminator drills holes into basement walls and slab foundations to inject the insecticide into the surrounding soil, rather than excavating it.

If you want to create a termite barrier yourself, you can dig a trench along the foundation. Dursban® (chlorpyrifos—Ortho-Klor® Soil Insect & Termite Killer) is available for use by homeowners and effectively controls termites. Apply the insecticide to the ditch and to the soil as you fill the trench back up. A sprinkling can is a handy applicator, but

---

Termite Inspections

Every 2 to 10 years, depending on how common termites are in your area, it is a good idea to hire a licensed pest control operator to inspect your home for termites. Before purchasing a home, be sure to have it inspected for termites; most banks and lending institutions require this before they will grant a mortgage.

A thorough inspection by a skilled professional can detect an infestation before it causes major damage, so the savings can be substantial. Many termite control companies also employ carpenters who can repair most types of termite damage, or can refer you to a carpenter. The inspection should be thorough and should include the attic. The inspector will also look for other wood-destroying insects and wood decay.

Making a yearly inspection yourself, as thoroughly as you can, may also result in a money-saving early detection of a problem. Be sure to check areas in which wood comes in contact with the ground, such as garage-door molding and porches.

If you have termites controlled professionally, it pays to get several quotes and to ask friends and neighbors for recommendations. Read the contract carefully; different companies may guarantee their work for differing periods of time. You may also elect to pay for an annual inspection after treatment; in this case any reinfestation in the treated area will be controlled at no charge. However, if the pest control operator has done a thorough job, this will rarely be necessary.

don't reuse it for anything but pesticides. Dig the trench 6 inches wide and down as far as is practical. Never dig below the top of the footing, or you may cause the foundation to shift. Use a layer of untreated soil at the very top when you fill the trench.

Follow label directions as to the correct amount of pesticide to apply per linear foot of foundation. Don't try to use less than the amount recommended—it may not prove effective.

Always precede such an application with a thorough inspection to determine the extent of the infestation. You may get effective control by treating only one large section of a foundation rather than the entire length. Follow all label precautions when applying the insecticide, including having good ventilation if you are working in a crawl space. Do not apply this insecticide near wells or ponds.

When properly applied, an insecticide can control termites for many years. There is no reason to make a routine second treatment. Reapply insecticide only if digging or some other activity has disrupted the barrier or if you see evidence of a reinfestation. In these cases, spot-treat only those areas.

Slab foundations are not invulnerable to termite attack, as was once thought. Because termites can get through an opening only 1/64 inch wide, they can easily penetrate the small cracks that concrete develops over the years. They may also enter other openings, such as those around pipes, at expansion joints, and behind bricks.

Treating under a slab means drilling holes at intervals and injecting the insecticide under pressure through the openings. Because this kind of treatment requires special skills and equipment, contact a licensed termite operator if it needs to be done.

If you are building in an area where subterranean termites are common, you would be well advised to pretreat the soil under the foundation for termites. A treatment at this time is much less expensive to make, and it may save you from a major repair expense later. Pretreatments are especially valuable before you pour any concrete slab or floor, because it is difficult to treat under them once they are laid. Treat the entire area under the slab, paying particular attention to such critical areas as where pipes, electrical conduits, and expansion joints penetrate.

Dampwood termites are simpler to control than subterranean termites. Because they must maintain contact with damp wood, simply eliminating moisture can usually control them. You may need to remove the damaged wood and replace it with pressure-treated lumber if you cannot control the moisture. Prevent any contact between wood and soil. You usually do not need to treat the soil with an insecticide, as you do for subterranean termites.

Controlling drywood termites can be very expensive. If the infestation is very small, replacing damaged boards or having a termite operator inject insecticides directly into the galleries may do the trick. But it can be very difficult to locate the entire colony, so having a termite operator fumigate the entire building is frequently the most reliable way to deal with these termites.

Drywood termites can penetrate flat surfaces of wood, but they much prefer to enter tight, narrow places before beginning their tunneling. Therefore, sealing any cracks and joints with putty and then painting over them helps stop these termites from entering. This, together with maintaining a sound coat of paint on all wood surfaces, reduces the chance that drywood termites will infest your home.

It is a good idea to remove any wood that has colonies of drywood termites, or any other type of termite, near your home. Firewood, scrap lumber, and dead branches of trees and shrubs are likely sources of drywood termites.

WOODWASPS (HORNTAILS)

Woodwasps, also called horntails, sometimes bore holes 1/6 to 1/4 inch in diameter in new homes as they escape from infested lumber. They actually attack only trees that are weak or were recently killed, such as those that have been logged or subject to a forest fire or smog stress, but these trees are often made into lumber used in home construction. You may see woodwasp exit holes in plaster, plasterboard, or paneling; the woodwasps actually developed in the lumber used for wall studs or subflooring. They escape by boring out through whatever material is covering this wood.

Fortunately, woodwasps do not harm household goods after they emerge, and they do not reinfest wood in a building. Often, people notice the holes without ever seeing the insects that made them.

The Pest
These large wasps measure about 1 inch long and are usually black or dark, metallic blue. They may fly to a window and find a way outdoors before you notice them. Woodwasps can't sting, but the females have very long, taillike projections that certainly look as if they could. This is actually an egg-laying device, known as an ovipositor. The wasps use it solely to insert eggs deep inside trees.

The cream-colored or yellowish white larvae eat the wood as they tunnel through it, creating circular holes. They form pupae inside the wood. The entire life cycle often takes two to three years, so adults may emerge long after the wood containing the larvae has been used in construction.

Prevention and Control
There is no practical method to prevent woodwasps from emerging. Tent fumigation of the entire home has not always been effective, and it is a very expensive and probably excessive treatment. If you apply an insecticide to the surface of the wood, adults can still bore out of it.

Because woodwasps tend to be few in number, and because they do not reinfest wood in structures, control measures are rarely necessary.

Woodwasps rarely make their nests in milled lumber. More commonly they hatch in lumber cut from already-infested trees.

In many cases the most practical thing to do is simply to repair the damage by sealing up the holes after the adults have emerged.

For future construction, choose wood that has been kiln-dried or vacuum-fumigated. Because these treatments make wood more expensive, low-value wood, such as studs and subflooring, is usually left untreated.

RODENTS AND OTHER ANIMAL PESTS

House mouse

Animals you might enjoy in the wild are not welcome when they move into your home. They can get into food, damage structures, and even carry diseases. Here are the best ways to remove them and keep them out.

Few animals are more universally hated and feared than rats, and with good reason. They are destructive, can spread disease, and cause fires by gnawing through wiring. Rat pests are of Asian origin but have spread throughout the world. Tropical cites have such huge rat infestations that health authorities estimate there to be one rat for every person. In most cities in the United States there is one rat for every 15 to 35 people.

Bats, too, are scary creatures. They are associated with witchcraft and evil only in Western culture, however. In China, for instance, they are considered a good omen and are welcomed in homes as symbols of good luck. Bats, though, should be discouraged around homes because they can be infected with rabies and their droppings spread disease organisms.

Most other animals are loved when encountered in the wild but become problems when they nest near or in homes. Wild animals belong in the outdoors, and they usually remain there. Occasionally, however, wild animals, including some birds, become household pests when they seek shelter in or around people's homes. Like insect pests, the most troublesome vertebrate pests tend to be those that thrive in close association with people. These pests include house mice, several rat species, and bats. Certain bats and birds that have become pests find buildings suitable as nesting sites. They are able to find enough food in cities, on farms, or in other human environments, and so they flourish wherever people live.

Raccoons can be pests in the country, suburbs, and even large cities. Their little masked faces startle people investigating a noise late at night.

Compared to insect pests, vertebrate pests are quite large. It can take only one to cause a significant problem or severe damage in a short amount of time. Animals eat more food than insects, and some gnaw on wood and even metal to gain entrance to houses.

Besides being destructive, animal pests can transmit diseases like rabies or rat-bite fever. Wild mammals and birds may bring fleas, ticks, mites, lice, and other companion pests that can infest homes. In some cases just their droppings present a significant nuisance and health problem.

Wild mammals and birds living near the house are usually best left alone unless they are actually nesting or taking shelter in the home or causing damage. The best way to keep wild animals from becoming pests is to deny them entry. This chapter provides information on modifying the house and other structures to make them inaccessible to or unsuitable for animal pests. A major preventative step is to screen all chimneys and vents and seal up all cracks and crevices around doors, windows, and siding. Rats and mice can actually unhinge and flatten their skulls, allowing them to squeeze through incredibly small openings. Preventing these animals from ever getting inside your home keeps them from becoming pests.

You will also want to be sure that you deny wild animals food. Raccoons may be cute, but the mess they can make is not. If you keep the garbage contained so that large mammals and rodents can't get at it, you will discourage them from visiting and deciding to stay.

When wild animals, including birds, do get into your home, the control methods differ significantly from those for insect pests. The emphasis is often on trapping the animals rather than eradicating them with baits and pesticides. Many people feel it is more humane to trap animals live and transport them to rural areas. This is unfortunately not always the case. Since animals are often strongly territorial, the transported animal may be attacked by competing animals in the new environment. Also, moving animals from one locality to another can spread rabies and other wildlife diseases. Check with the local wildlife authority before trapping an animal live, because trapping or transporting certain wild animals may be against the law in your state.

BATS

Because they fly at night, bats usually go unnoticed; you may have them in your yard or neighborhood without realizing it. Bats feed on insects, devouring as many as 3,000 mosquitoes or other flying insects every night. Most bats roost outdoors in dark, secluded places, such as in caves, in hollow trees, and under loose bark. They become a problem, however, when they choose buildings as roosts. Attics and church steeples are favorite sites, if bats can find an entrance. A bat can squeeze through an opening as small as ⅜ inch.

Bat droppings may accumulate in buildings, attracting insects and resulting in an objectionable odor. Bat squeaks and the rustling noise they make as they enter or leave their roosts can also be bothersome.

All North American species of bat feed on insects. Only the vampire bat of Central and South America feeds on mammal blood; it can injure livestock and transmit rabies. Because about 1 in every 1,000 bats carries rabies,

anyone bitten or scratched by a bat should see a physician in case a postexposure rabies immunization is needed.

Although only a small percentage of bats carry rabies, those that do are more likely to come into direct contact with humans than are healthy bats. Rabies-infected bats rarely exhibit the symptoms of rabid dogs, but the disease affects them in other ways. Infected bats are more likely to appear during daylight hours than are healthy bats, and they may be weak and unable to cling to their roosts. Because of this, infected bats may flop around helplessly near the entrance to their roosts.

Most bat bites occur when someone tries to pick up a weak bat. Rabies can be transmitted to humans by the bite or scratch of an infected bat. Unless treated by a physician, rabies is usually fatal to humans and other mammals. If you find a sick or dead bat, notify local public health officials immediately; don't try to handle it yourself.

In case of a bat bite, wash the area immediately with hot, soapy water, and consult a

Although bats do not usually attack humans, a sick bat may bite if disturbed. Since bats may carry rabies, it is best never to touch them.

physician promptly. If possible, capture the bat without damaging the head, place it in a jar or plastic bag, and keep it refrigerated. Health authorities may want the bat for a rabies test.

Bats also carry a variety of insects and mites, particularly the bat bug, that live on the outside of their bodies. This insect is related to the common bed bug. It will bite people, and it sometimes becomes a problem in bat-infested homes.

Bat guano (droppings) can attract cockroaches, flies, and similar pests. It also is a breeding place for the fungus *Histoplasma capsulatum*. The airborne spores of this fungus cause a lung disease. Because of this fungus, anyone removing bat droppings should wear a respirator and protective clothing.

The Pest

Bats represent a unique and interesting group of animals because they are the only mammals that can truly fly. Thin membranes of bare or slightly hairy skin stretched between their front and hind legs create the wings. Bats' bodies are only a few inches long, but their wingspans can be from 8 to 14 inches, depending on the species.

Bats fly at night, emitting ultrasonic sounds to help them detect and track flying insects. When a bat gets close to an insect, it scoops it up

with its wings or short tail and stuffs the insect into its mouth. A bat can eat approximately half its body weight in insects each night.

Bats, like other mammals, give birth to live young and nurse them with milk. The young are generally born during late spring and early summer. Within a month, the young make feeding flights. By fall, even young bats can fly long distances. Bats can live 10 to 20 years, depending on the species.

Some bat species migrate to warmer areas for the winter rather than spending it in buildings. They tend to return to the same roosting site each year. Most species roost alone and seldom roost in buildings. The species that live in large colonies are the ones that can become noticeable nuisances and health problems.

Prevention and Control

Because bats eat large quantities of insects, they should not be killed needlessly. In most cases where bats are infesting buildings, the problem can be eliminated by excluding the bats. Only rarely is it necessary to kill them. Individual, apparently sick bats should be destroyed by public health authorities or a professional pest control operator.

The only permanent way to get rid of healthy bats that are roosting in an attic or other building is to wait until they leave for their evening flight and then block their entry points. Bats are sometimes attracted to a roosting site by odors left by a previous colony and are thus likely to reinvade the structure unless you block the openings.

To determine the bats' entry and exit places, watch for emerging bats from about one-half hour before dusk to about an hour after the first bat emerges. Common entry points include openings beneath eaves, the joint between a chimney and the house, the entry points of pipes and electrical lines, and the spaces

Wait until early evening, when bats have flown for the night, to plug attic openings. Unplug at least one opening the following evening to be sure all bats have had a chance to escape.

behind loose shingles, siding, and vents.

After you locate the entry places, make the needed repairs to seal them up. Unlike rats and mice, bats cannot gnaw through wood or other building materials. You can use anything that will block the openings, including wood, caulk, weather stripping, and ¼-inch-mesh hardware cloth.

Seal the openings after the bats leave the roost at twilight or after they disperse in the fall so that you do not trap any bats inside. If any die inside the structure, their carcasses may cause an odor. Batproofing is not recommended from mid-May to mid-August because new-born, flightless bats may be present in the roosts. Unplug the major openings early the next evening to allow any remaining bats to escape, and seal them again before the bats return later that night. Repeat this procedure if you hear or see any remaining bats within the structure. Watch the building for several evenings to see if there are any openings you might have overlooked.

In buildings where it is not possible to close all openings, try setting up bright lights in the roost to provide constant illumination day and night. You can also increase the air circulation with an electric fan; bats dislike strong air currents where they roost. These techniques have been effective in some cases but will fail if the bats can escape from the light or breeze. Mothballs, crystals, or flakes may discourage bats in confined areas where the vapors can build up. Three to five pounds should be adequate for an average attic. Place them in open boxes or in old stockings so that you can easily remove them if you find the odor too strong or if they don't work.

Occasionally, a bat accidentally finds its way into the living area of a home. When this occurs, the best strategy is to encourage it to leave by opening all doors and windows. Bats easily detect slight air currents coming through openings and will often fly out. If it is dark outside, turn off all lights; otherwise, the bat may seek a dark refuge behind a curtain or in some other location. You can use a broom to dislodge a bat from a dark spot.

As a last resort, try to catch a flying bat with a net or in a wicker basket. Even though bats are blind, their sonar allows them to detect and avoid solid containers, such as metal wastebaskets and plastic pails; they may be confused by a wicker wastebasket or other insubstantial object, however. Hold the basket in the bat's flight path. Release the trapped bat outdoors. Do not handle a bat unless you are wearing heavy gloves.

BIRDS

Most birds are a welcome sight around homes and gardens. They have cheerful songs, are entertaining to watch and study, and many consume huge numbers of insects. However, a few bird species can cause serious problems. Three of the worst bird pests—house sparrows, pigeons, and starlings—have adapted especially well to human environments and are common problems in cities and suburbs.

Some nuisance birds build their nests in chimneys, drains, and air-intake vents, causing them to become clogged. The droppings from nests and roosting sites can dirty sidewalks and window ledges as well as deface statues. Because of their acidity, bird droppings accelerate the deterioration of almost anything they land on. Also, some bird noises, such as a woodpecker's drumming activity or the clatter of huge flocks of starlings, are disagreeable or distracting.

Pigeons may be fun to feed in the park, but their roosting sites are dirty and expose people to many diseases.

Birds are occasionally a health hazard because they can carry diseases, such as encephalitis and toxoplasmosis, that affect people. Bird droppings can be breeding places for salmonella bacteria and, like bat droppings, the fungus that causes histoplasmosis. Birds may also carry a number of bugs, fleas, ticks, and mites, some of which bite humans and pets. When birds nest in or on houses, their nests may be breeding sites for fabric pests, such as carpet beetles and clothes moths, and certain pantry pests, all of which may invade the house after the birds leave.

The Pest

A number of birds can be nuisances around homes. Cliff swallows build mud nests underneath eaves and in similar sites. The nests are shaped like gourds, with a round, necklike entrance. These migratory birds overwinter in South America. Their nesting period lasts from March through June, during which time the area beneath the nests is bombarded with dropped mud and feces. Nests are usually grouped together in colonies and located near a supply of water. Cliff swallows use the same nesting sites year after year, leaving the area by the end of June. They feed almost exclusively on insects.

House sparrows, also called English sparrows, are not really sparrows but belong to the weaver finch family. They were introduced into New York from England in 1850 and have since spread across most of the continent. These highly social birds often build their nests close together, choosing somewhat protected, elevated places such as rain gutters and attic rafters. The messy nests are composed of grass, straw, and other debris. House sparrows have adapted extremely well to human environments and are most abundant in cities and around farm buildings.

House sparrows are prolific breeders, raising from two to five broods per year and averaging about five eggs per clutch. The adults feed primarily on grain and other seeds, but feed mostly insects to their young.

Pigeons were introduced into North America by French settlers in Quebec about 1606. Although they are a favorite bird for many people in parks, when pigeons become abundant they can be serious pests. Their droppings are messy and carry an unusually large number of diseases, perhaps more than any other bird species.

Pigeons commonly build nests on ledges of buildings. Their nests consist of twigs and grasses put together loosely. Pigeons are monogamous and mate for life. They breed the year around, with the peak reproduction period occurring in spring and summer. They lay a clutch of one or two eggs several times a year. In captivity pigeons can live for 30 years or more, but in the wild most apparently live less than five years. Pigeons feed primarily on seeds and grain augmented by some fruits, vegetables, and insects.

Starlings were deliberately introduced into New York from England in the 1890s and have since spread to all 50 states. They mate in the spring and build nests in tree hollows, abandoned woodpecker holes, and sometimes in crevices on buildings. The nests are composed of fine grass and straw and other soft materials. When starlings aren't nesting, they form large, communal roosts. They group together in flocks that grow as the summer progresses until they consist of hundreds or thousands of birds. At dusk these huge flocks may come into a city and rest on building ledges, billboards, park trees, and similar spots. Their copious droppings can be a messy and unsightly problem, and many people also dislike the birds' characteristic loud, harsh, squawking calls.

Starlings are omnivorous, feeding mainly on insects but also consuming berries and other fruits. They lay an average of five eggs per clutch and produce from one to three broods per year. In the fall starlings in northern areas tend to move to warmer areas in the South.

Woodpeckers and flickers cause two kinds of problems—noise and pecking damage. The noise results when males proclaim their territory by "drumming." They peck at nearly anything that will produce a loud sound, including hollow trees, metal pipes and siding, and glass. When they do this on a house, the racket can be horrible. They often drum in the early morning hours, especially in the spring and early summer.

Other damage may result when woodpeckers drill into wooden structures, either to make nesting sites or to feed on insects beneath the surface of the wood. Acorn woodpeckers drill holes that are large enough to accommodate an acorn. Most woodpeckers feed primarily on wood-boring insects and other insects beneath the surface of wood, in bark cracks, or under loose bark.

Cliff swallows live in colonies. They have been known to build hundreds of nests on the sides of buildings or bridges.

Woodpeckers are most common in wooded areas. Some species migrate each year, but most spend their lives around a particular patch of woods. Because they are territorial, damage to a building often results from only one or two birds.

Other birds occasionally become a problem. Barn swallows build messy nests on rafters inside barns, sheds, and other buildings. Chimney swifts build nests in sheltered spots high up on buildings, and sometimes in chimneys. They return to the same nest year after year and enlarge it. Blackbirds, cowbirds, and grackles sometimes combine with starlings to form large roosting populations.

Prevention and Control

Because birds are for the most part considered desirable, federal and local laws may determine how you can deal with them. Cliff swallows and woodpeckers are protected by the Federal Bird Treaty Act, which makes it illegal to kill them. Birds not protected by federal laws include starlings, house sparrows, and common pigeons. These birds generally can be killed if they are damaging a home or garden. However, in some communities local laws may protect the birds or regulate the control methods you can use. Contact a county wildlife officer if you are in doubt about local laws regarding a specific bird.

There are no laws against scaring or excluding birds that are damaging your home or garden. You can try to frighten birds with fake snakes or owls or with flashing shiny objects such as aluminum foil or strips of mylar. Loud noises such as carbide exploders or firecrackers can also be effective. Most of these methods work only briefly, because birds are able to adjust to these new sights and sounds.

It is best to frighten birds away on their first few visits, because once they become accustomed to an area they will be more difficult to frighten.

Recordings of starling distress calls have been effective in getting large flocks of starlings to move their roosting site. You should start using the recordings as soon as the birds begin roosting at the site. The sounds must be started in the early evening, when the birds first begin to arrive, and they may have to be played for several evenings.

Excluding birds from a spot may be a good solution if they are nesting in an undesirable location. You can do this by stringing bird netting or rust-proof wire netting such as chicken wire over the spot. You can also apply a sticky bird repellent or string lines above a ledge so that the birds can't land. Some pest control operators install specially manufactured strips of tiny, sharp projections that the birds avoid. This treatment is a long-term solution but can be fairly expensive. If the birds are entering a building to nest, sealing the openings may be all that is needed.

You can discourage birds such as pigeons from roosting on the house by stretching rustproof wire netting over ledges and eaves.

Live bird traps are available but present the problem of what to do with the trapped birds. Driving them out to the country and releasing them is useless; they will simply return.

No lethal chemicals are registered for homeowners to use on birds. Depending on local laws, however, pest control operators may be able to use various avicides. The chemicals are either applied to perches to kill birds resting on them or mixed into a bait. The poison can kill nonpest birds if not used carefully. If dead birds aren't picked up immediately, animals may eat them and also be poisoned. Because of these problems, avicides should be used only by professionals who know the habits of birds.

You can prune trees to open up their canopies to discourage starlings and other communal roosting birds. These birds prefer to roost in trees with dense rather than open canopies.

Before attempting to remove cliff or barn swallow nests, you'll need a depredation permit from the Fish and Wildlife Service. You can discourage swallows by washing down their nests with a strong stream of water before they lay eggs or after the baby birds mature. You will probably need to do this daily because the birds are persistent in rebuilding their nests. To prevent them from rebuilding after you have removed the nests, block their access to the site by stringing up bird netting or chicken wire.

You can prevent woodpecker drumming and damage by stringing nylon bird netting or chicken wire over the spot or by applying a sticky bird repellent. Some sticky material stains wood, however; test it on a small area first. Removing the object the woodpecker is drumming on is another remedy.

If insect pests breeding in bird nests are a problem, control them by dusting the nest with a Sevin® (carbaryl) dust.

MICE AND RATS

The mere sight of a mouse or a rat can cause a tremendous commotion. These common household pests are universally disliked and feared. Although their presence doesn't necessarily reflect on household standards, mice and rats can bring disease and filth with them.

Mice and rats ruin fabrics when they gather nesting materials. Both destroy food by eating it and contaminating it with their feces. They even gnaw on wiring, causing household fires and appliance breakdowns. Odors from their droppings, urine, nests, or carcasses are further problems.

Mice and rats can transmit a wide variety of diseases. Their droppings sometimes contain salmonella bacteria, which can cause food poisoning if they contaminate food. Rodent bites can transmit rat-bite fever. Other rodent-borne diseases include leptospirosis, murine typhus, rickettsial pox, lymphocytic choriomeningitis, and bubonic plague. Fortunately, these diseases are not common, but the possibility that they may occur in the area is reason enough to control rodents.

The Pests

House mice originated in central Asia but now live throughout the world. They eat the same kinds of foods as humans, generally preferring cereal grains but also eating sweets, meat, and nuts. Mice obtain their daily intake by nibbling on food a little bit at a time rather than consuming a large amount at once and in one place. Although they will drink water, they can obtain enough moisture from the food they eat.

Even the largest adult house mouse weighs only about 1 ounce or less and is shorter than 3¾ inches, not counting the tail. Because of their small size, mice are able to enter a crack only ¼ inch wide, making it a challenge to

Above: The Norway rat is the most common rat in the United States and is often found in sewer systems. Note that the tail is shorter than the head and the body.
Below: House mice are cute, but they multiply quickly and make it very difficult to keep a clean and safe food supply.

keep them out of the house. House mice are light gray to dark brown with lighter undersides. They normally live about one year. Each female produces as many as 6 to 10 litters per year, with an average of about 6 to 8 young in each litter. They make nests of shredded cloth, paper, or other soft material. Breeding may slow down or stop at very high population densities. Mice living indoors breed throughout the year. Breeding among mice living outdoors tends to peak in the spring and fall. Outdoor mice often move indoors when the weather starts to grow cool in the fall.

Mice remain within a very small territory. They may spend their lives within just a few yards if they have enough food and shelter there. Each night they travel over their territories, carefully investigating any new object or any change in position of a familiar object. They follow regular pathways, usually along walls and behind or under objects rather than in the open.

Deer mice sometimes invade houses and cabins in fields and woodlands. These natives are about the same size as house mice, but they have larger ears and eyes and distinctly white undersides. Their tails are bicolored with well-defined lines where the white ends and the brown or gray uppersides begin. Native meadow mice (voles) sometimes enter ground-floor rooms but almost never become established and reproduce. They are most common in areas with dense grass cover. Adults are about 5 inches long. Although the biology and habits of these mice differ, control measures are similar.

Two rat species—the Norway rat and the roof rat—are pests in North America. Norway rats, the most common species, occur throughout the United States. Roof rats, also known as ship rats or black rats, have various color phases and are sometimes referred to as white-bellied or gray-bellied rats. They occur primarily in eastern and Gulf Coast regions and the Pacific Coast states.

Norway rats often build their nests in burrows leading under buildings, in piles of debris, or in stream banks. The burrows measure 3 to 5 inches in diameter and have secondary escape holes that are hidden under grass, boards, or other coverings. They also nest in hidden areas under buildings.

Norway rats are gray to brown and are 7½ to 10 inches long, not including their hairless, scaly tails. They normally live from 9 to 12 months in the wild and can produce four to seven litters, each with 8 to 12 young. In cold-winter climates, peak breeding occurs in the spring and fall.

Adult roof rats are 6 to 8½ inches long. They live 9 to 12 months in the wild and have about six litters, each with six to eight young. Roof rats can live in the same types of locations as Norway rats, but they are better climbers and are much more likely to build their nests in walls, attics, vines, or trees. Common nesting sites include ivy and, in very warm climates, palm trees that haven't had their old fronds removed. Their great agility enables them to move along telephone and utility lines, as well as to climb up vertical pipes, with ease.

Rats are omnivorous, eating any food people eat. They can get all the water they need by eating fruits, vegetables, and other high-moisture foods, but if only dry food is available they must get water from leaky plumbing, sewer pipes, heavy condensation, and similar sources. Rats sometimes chew through plastic pipes to obtain water.

Like mice, rats are active primarily at night and use regular pathways. They have a much larger range than mice, and unlike mice they tend to be very cautious about any new food or other item in their territory.

Prevention and Control

For best results in controlling mice and rats, general sanitation (reducing the food, water, and shelter available to pests) should always be combined with other rodent control measures. In general, good sanitation works better to help control rats than mice because mice can live on relatively small amounts of food. Besides making it more difficult for these rodents to live and reproduce, good sanitation increases the likelihood that they will eat a poisoned bait or a bait in a trap.

Sanitation measures include cleaning up all spills and crumbs and keeping a tight lid on both outdoor garbage cans and indoor kitchen waste cans. All opened food should be kept in the refrigerator or in containers with tight-fitting lids. Do not leave pet food out in dishes overnight. (See pages 11 and 12 for more details on sanitation.)

You can also reduce possible rodent shelters outside the house by removing nearby piles of loose boards and other debris. For roof rats in particular, it is important to remove dense, overgrown ornamental trees and shrubs that may serve as nesting sites. Because roof rats are good climbers and can easily climb up trees, prune back tree branches so that they are several feet from the house.

One of the most important steps in rodentproofing your home is to seal properly all possible openings through which rodents might enter. Because rats and mice that live outdoors tend to migrate indoors in the fall, a good practice is to undertake a thorough annual inspection of the house in late summer or early fall. Make any repairs or modifications necessary to keep these pests out.

House mice and Norway rats generally enter openings near the foundation or several feet above the ground. If you have roof rats in the area, your inspection should also include areas high up on the house near the roofline and on the roof. Seal all cracks and openings that are ¼ inch or larger. Check for openings where pipes, wires, and vents enter the house. All ventilation screens in the foundation and attic should fit tightly to exclude rodents. Use weather stripping under and around garage doors and other doors so that they fit tightly. (See pages 7 to 11 for additional suggestions on ways you can pestproof your home.)

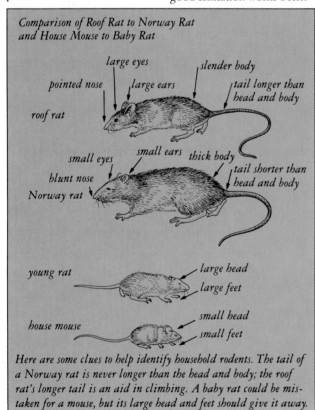

Comparison of Roof Rat to Norway Rat and House Mouse to Baby Rat

large eyes — pointed nose — large ears — slender body — tail longer than head and body — roof rat

small eyes — small ears — thick body — blunt nose — tail shorter than head and body — Norway rat

young rat — large head — large feet

house mouse — small head — small feet

Here are some clues to help identify household rodents. The tail of a Norway rat is never longer than the head and body; the roof rat's longer tail is an aid in climbing. A baby rat could be mistaken for a mouse, but its large head and feet should give it away.

Check your home carefully for openings through which rodents could enter. Make sure that vents are covered with sound and tightly fitted screens. Remember that mice are able to crawl through a crack that is only ¼ inch wide.

Materials you can use to seal holes include steel wool (pushed tightly into holes), wire screen, sheet metal, and cement. In general, stay away from wood, plastic sheeting, and similar materials, because rodents are able to gnaw through them.

Norway rats in particular often infest sewer systems. In old cities or older sections of cities, they often come above ground through breaks in underground sewer lines. To resolve the problem, pipes need to be excavated and repairs made. Be sure that all floor drains are properly capped so that rats can't come from the sewer lines. It may also be necessary to screen large sewer air vents that extend to the roof. Where sewer systems are heavily infested, rats may enter a house through toilets, although this is rare.

Mice are extremely capable when it comes to getting into buildings, and even the most conscientious rodentproofing measures may not keep your home completely free of these pests. Trapping and baiting are the best ways to get rid of mice and rats that are already indoors. Both techniques are effective. If the infestation is large, baiting is the quickest way to bring it under control. If you are using one technique and it doesn't seem to be working, the pests may have become either trap-shy or bait-shy. If this happens, try switching to the other control method. You must place baits and traps within the rodents' territories to catch them.

The wooden snap trap, sometimes called the breakback or spring trap, is the most common trap used today. Invented around 1895, it has outlived thousands of other mousetrap designs.

To improve your catches with the snap trap, be sure to use the correct size trap, bait it properly, and place it correctly. It is important to choose the correct size for the pests you have; mice usually

escape unharmed from rat-traps because the metal bar misses them completely. Mousetraps are too small for adult rats, although they can catch very young rats. Encountering the wrong size snap trap is one way that these pests become trap-shy.

You can usually increase your catch by first placing unset but baited traps in the pests' territories for several days. This is particularly important for rats but may also improve mouse control. Check the traps daily, and replace the bait. Once the rodent is used to the trap, set it.

The kind of bait you use is also important. Nut meats, dried fruits, and bacon make excellent baits for mice and rats, but a large variety of other foods also work successfully, including peanut butter, gumdrops, and cheese. The best baits are sometimes items the pest has already been eating.

If you prebait, you can also use this occasion to test different baits. Choose the item they seem to prefer as your final bait. When you set the trap, tie the bait to the trigger to make it more difficult to

remove so that the pest will be sure to trip the catch. If you use a soft bait, such as peanut butter, push it firmly onto the trigger.

Place the traps in natural pathways, such as along walls, behind objects, and in dark corners, to increase the chance that the pests will encounter them. A good place to set traps is where you see droppings or other signs of rodent presence. For roof rats, you can wire traps to tree limbs, overhead pipes, beams, or other paths.

Do not set a trap parallel to the wall unless you use two traps. In this event, set them end to end against the wall so that the triggers face away from each other. This way, the rodent will spring the trap regardless of the direction in which it is traveling. If you use one trap, place it perpendicular to the wall, with the trigger toward the wall. This improves the chance that the pest will pass over the trigger.

For mice, space the traps 10 feet or less apart. Move them to a new spot if you are unsuccessful after four or five days. Because rats have a

wider range, you can place traps farther apart. Use a large number of traps, and set them all at once. This helps prevent mice and rats from becoming trap-shy and greatly improves your success. A good rule of thumb is to use one to three traps for every rodent you think you have. For a heavy infestation, you may need a couple dozen traps.

Check the traps daily. Wear plastic gloves when you remove the mouse or rat, and place it in a plastic bag. Seal the bag and discard it in the garbage, or bury it.

One trap design that has become more popular recently is the sticky or glue trap. These were designed for use in warehouses and similar settings and have limited use in homes. Pets can get stuck to them, and the glue also adheres to rugs and shoes. When you do catch a rodent, it remains alive, and disposal thus becomes a problem. Sticky rodent traps only rarely catch rodents that could not just as well be caught with a snap trap.

Always place traps, especially rattraps, in places where children and pets can't get to them. If this is not possible, you may need to build boxes with holes in the ends. Place the traps inside the boxes, and secure the lids.

Poisoned baits effectively kill rodents that eat them, and a variety are available for mouse and rat control. Not all mouse baits will control rats, and vice versa, so be sure to use a bait labeled for the kind of rodent you have.

The rodenticides in these baits can also kill larger animals, but they are safe if used as directed. Anticoagulants such as warfarin, diphacinone, and brodifacoum have several disadvantages. They act as a blood anticoagulant and cause the animal to die slowly. Rodents eating anticoagulant bait often continue to feed even after ingesting a lethal dose. If they do this, their dead carcasses may pose

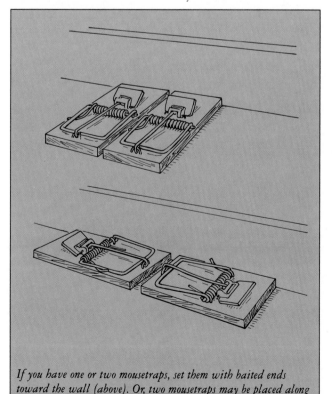

If you have one or two mousetraps, set them with baited ends toward the wall (above). Or, two mousetraps may be placed along the wall with baited ends in opposite directions (below).

a hazard to wildlife and pets that may feed on them. In some areas, Norway and roof rats are resistant to warfarin and other anticoagulant baits. Warfarin resistance in house mice may be even more widespread. Where resistance is a problem, cholecalciferol baits are an effective method of control.

Mouse and rat baits containing cholecalciferol (Ortho Rat-B-Gon® Rat & Mouse Killer Bait and Ortho Mouse-B-Gon® Mouse Killer) are among the safest baits to use around homes. Cholecalciferol is a unique rodenticide that kills rodents within 2 to 3 days by elevating the blood level of calcium to a fatal level, causing a heart attack. Rodents are very sensitive to changes in blood calcium, and because of their small size, they die from consuming very small amounts of cholecalciferol. Because these baits have a low concentration of rodenticide they are safer to use around pets and desirable wild animals.

Often a single feeding is sufficient to kill the rodent, but mice may have to consume the bait over several consecutive days since they tend to nibble, not eating large quantities at one time.

Even though many baits aren't immediately toxic to people or pets, always place them in areas that are inaccessible to children and domestic animals. If you can't do so, nail boards over the bait station or build bait boxes with holes in them and a lid that you can secure. Each box should have two 1-inch openings for mice or two 2½-inch openings for rats. Make the box large enough to accommodate several rats or mice at one time.

Place baits in the same kinds of places as traps. Good locations include pathways, burrows, and hiding places. For roof rats, place the bait up high, such as in the crotch of a tree, on the top of a fence, or high in a vine; tie it so that it will not fall to the ground. Make sure that the bait will stay dry if it rains.

Leave the bait in place so that the rodents can feed on it continuously for at least 10 days or until all signs of activity cease. Replenish the bait as needed so that there is always a supply during the baiting period. Discard any bait that becomes moldy or old, being careful to follow all label directions for proper disposal.

Rodents killed by bait may die near the bait or return to their nests and die; be sure to dispose of dead carcasses properly. Occasionally, an animal will die in an inaccessible part of the house. The odor from a dead rat can cause a stench for a few weeks. Mice are small enough that the odor from a decaying carcass often can't be detected.

Besides trapping or baiting rodents, there are few other effective ways to control them. Manufacturers of electromagnetic and ultrasonic devices claim that these devices help repel or kill rodents; repeated tests at various universities have found them to be ineffective, however. Some ultrasonic devices can

repel rodents to a small degree but rapidly become useless as the rodents become accustomed to the high-frequency sounds.

Cats are time-honored, natural rodent controllers. Cats that are good hunters can be helpful in keeping rodent populations around farms under control, but pet cats are usually not sufficient around homes because they didn't learn to hunt while they were young.

OTHER HOUSEHOLD ANIMAL PESTS

The animals discussed in this section—opossums, raccoons, porcupines, skunks, tree squirrels, and chipmunks—are only occasional pests. They become bothersome when they take up residence or build a nest in or under your home. Treat all of these animals cautiously, because they may try to bite if you corner them. And keep in mind that in some areas skunks and raccoons carry rabies. Another reason to discourage wild animals from nesting under your house is that fleas, lice, and mites may accompany them. These can sometimes become a household problem when the animal leaves or dies.

The Pests

Opossums sometimes get into crawl spaces, basements, and even attics. These sluggish animals are omnivorous and will raid garbage cans. Opossums are the only marsupial native to North America. When born, they are the size of a small bumblebee and quickly crawl to their mother's pouch, where they nurse for about three months.

Porcupines occasionally build their dens under buildings. They also gnaw on almost any exposed wood, especially wooden tool handles

Signs of Mice and Rats

smudges

gnaw marks

tracks

droppings

You don't have to see rodents to know they are around. They leave many signs that give their presence away. The most important of these are their droppings and runways. One pair of mice deposits about 16,000 droppings in a six-month period. You'll see them nearly everywhere that the mice feed and along their runways. Mouse droppings are dark colored, about ¼ inch long, and as thick as the lead in a pencil. Rat droppings are considerably larger. The droppings are soft at first and become hard after several days. Rats and, to a lesser degree, mice leave grease marks where their bodies continuously rub against walls along well-traveled runways. Other signs of rodents include gnawed wood or food, burrows, shredded paper, tracks, or excited pets. You may also smell a rat's musky odor or hear running or gnawing.

You can rent live traps in many areas, but be sure to consult local wildlife authorities for advice on what to do with trapped animals. Here an opossum has been captured.

that are salty with sweat. The adults are covered with about 20,000 sharp, barbed quills that they use for defense. Contrary to popular belief, porcupines cannot shoot their quills. The animal must make direct contact with an attacker, either by slapping it with its tail or lunging at it. The quills easily come off the porcupine, but their barbs make them hard to extract from the victim's skin.

Raccoons are notorious for tipping over garbage cans, creating a mess and commotion in the process. They also devour ornamental fish in garden ponds. They occasionally nest in attics, chimneys, crawl spaces, and other secluded areas. Raccoons can be very dangerous if they are cornered and can kill some breeds of dog.

Skunks sometimes find shelter beneath buildings. When disturbed, they may discharge their characteristic pungent scent; the family dog usually is the one that gets sprayed. To neutralize the skunk scent, use neutroleum alpha, available from hospital supply outlets. Rabid skunks are common in some areas.

Tree squirrels and chipmunks are pesty when they enter attics, garages, or chimneys to nest or store food. They are especially noisy and may gnaw electrical wiring, tear up building insulation, and destroy stored items, leaving quite a mess.

Prevention and Control

One of the best ways to deal with these pest animals is to trap and remove them. However, local laws may protect certain animals. Check with a local wildlife authority before beginning to trap.

Traps are of two types: those that kill the animal immediately and those that take it live. Because traps may catch dogs, cats, and other nonpest animals, live traps are the safest ones to use. Most are rectangular, with a door at one or both ends, and are made of wire mesh, sheet metal, or wood. Some hardware and garden stores sell them, and vector control and wildlife agencies sometimes will loan live traps out for a short period.

If you catch an animal with a live trap, you are faced with how to get rid of it. Many people feel that it is more humane to relocate the animal than to destroy it. This is not always a good idea, however. Since many animals are territorial, a relocated animal may have little chance of surviving. Because relocating wild animals may also spread wildlife diseases, this practice is prohibited in some areas. The local wildlife authority will be able to advise you.

Place a trap along suspected routes of travel or directly in the animal's trail. Put bait inside the trap, but also place a small amount outside the trap to entice it inside.

Traps must be properly baited to entice the animal inside. Opossums are not suspicious and will enter most

traps without hesitating. Good opossum baits include raw red meat, fish, chicken entrails, vegetables, and canned dog food. Porcupines are also easy to trap. Good porcupine baits include apples and salted carrots. For squirrels and chipmunks, use nuts, peanut butter, or raisins.

Raccoons are clever and strong and can be difficult to trap. You might increase your chances by camouflaging the trap with brush and twigs and/or by pushing the bottom of the trap into the soil so that the bottom wires are covered. Good baits include fish (especially smoked fish), fish-flavored cat food, chicken parts, honey-coated bread or ears of corn, and pieces of melon.

Good skunk baits are canned or fresh fish, canned fish-flavored cat food, and chicken parts. To trap a skunk, place the trap at the opening to its den or along its

When a skunk has taken up residence under or near your house, chances are great that you or your dog will accidentally cause a big stink.

pathway, such as along a fence or wall. Once you've caught it, be careful not to provoke it into ejecting its odor. It can shoot its scent as far as 10 feet. Approach the cage slowly so that you don't startle the animal, and cover it with a burlap bag or plastic garbage bag. Keeping it in the dark and handling it gently should do the trick.

Rat snap traps can catch squirrels, but in some localities squirrels are protected by law. To catch a particularly

Squirrels usually enter buildings through openings above eye level. Sheet metal or hardware cloth over vents helps keep them out of your home.

wary squirrel, put a baited trap in position but wait several days before setting the trip mechanism. For a live trap, keep additional bait near the entrance and keep the doors to the trap tied open.

Whenever you have set a trap for an animal, check it once, and preferably twice, daily. The morning is a good time because most of these animals are nocturnal. Never handle a live trapped animal, even with heavy gloves; trapped animals are irritable. The animals described here all have sharp teeth and strong jaws and may bite.

Mothballs or flakes may repel wild animals when placed in their burrows or other enclosed areas. Use 1 to 2 pounds in a burrow or 5 to 10 pounds in an average crawl space or attic.

If the animal is living in or under the house, find its entry and plug it up when the animal is outside or after you trap it. Use wire screening, sheet metal, or similar materials because these animals may gnaw through wood.

To keep animals out of the attic or chimney, it helps to prune off any tree limbs within 6 feet of the house. Cover the chimney with a protective cap and a hardware cloth screen. To discourage animals, keep garbage cans tightly closed and anchored in racks if need be. If any animal enters a garage or basement, open all doors to let it exit on its own accord. Do not prod or disturb it.

BITING AND STINGING PESTS

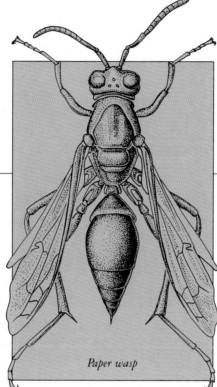

Paper wasp

The most feared pests are probably not those that damage household belongings but those that bite or sting. Biting and stinging insects and other arthropods, such as spiders and their relatives, cause pain and sickness either directly or by the diseases they transmit. Many of these pests—such as ants, bees, centipedes, wasps, scorpions, and spiders—sting simply to defend themselves or their colony from a real or perceived threat. If you don't threaten them, they will not harm you. Centipedes, scorpions, and spiders quickly retreat when a person is near and seldom sting unless someone accidentally touches or steps on them. By learning to identify these insects and by understanding a little about their habits you can avoid encounters with them.

Of the many millions of people stung by ants, bees, wasps, and hornets or bitten by venomous spiders and other arthropods each year in the United States, most feel only temporary pain. About 25,000 people have severe allergic reactions, but fewer than 100 people die each year as a direct result of these bites and stings. Some authorities believe the actual death rates may be higher than those reported and suspect that many sudden deaths attributed to heart attacks may actually have resulted from insect stings. Bites and stings from bees, wasps, and ants cause the vast majority of these fatalities, even though certain other pests produce much stronger venom. This is because these insects are more common and therefore sting more people than other species do.

Although they attack only to defend themselves or their hive, honey bees are such common insects that few people have escaped their sting.

Some insects and spiders found in your yard or home have painful stings or bites, which may cause allergic reactions. Proper control and prevention measures can eliminate these menaces from your environment.

Most of the deaths caused by bees, wasps, and ants are in people who are highly allergic to the venom. The victim suffers anaphylactic shock and dies if lifesaving procedures are not administered in time. The first recorded death from an allergic reaction was probably that of King Menes of Egypt, who died from a hornet sting in 2461 B.C. Only a small percentage of people are this sensitive; most people could survive even multiple stings from a colony defending itself.

Because an extreme anaphylactic reaction can kill a person within an hour, you should consult a physician if you are extremely allergic. You may be advised to carry an emergency kit or have desensitizing shots.

Most of the other pests described in this chapter sting or bite to obtain blood. These pests include various kinds of bugs, mites, fleas, flies, lice, mosquitoes, and ticks. Human lice are the only pests in this group that feed exclusively on people; the others feed on a wide variety of hosts and often prefer a host other than a human if they can find one. Applying an insect repellent to your skin and clothes is one of the best ways to defend yourself from these pests when you are outdoors in an infested area.

When homes become infested with pests such as fleas or ticks, insecticides can bring relief. But for control to be effective, the insecticide must be applied simultaneously to all infested areas—both indoors and out—and to pets to ensure complete control. If the outside source of the infestation isn't controlled, your home will continue to become reinfested.

Fortunately, most insect-transmitted diseases in the United States have either become less severe because of better medical treatments or less common due to better vector control. Malaria, for instance, affected over 100,000 people annually in the United States as recently as the 1930s; today, it no longer occurs. The eradication of malaria resulted from a better understanding of the disease-carrying mosquitoes and the development of effective control methods. Constant vigilance by public agencies once they had the proper tools was the key factor in eliminating malaria from the continental United States.

ANTS

The ants discussed here—fire ants, harvester ants, and field ants—are ones whose bites are especially painful. (Common household ants are discussed on page 32, and carpenter ants are covered on page 54.) Although no ant sting causes as much pain as a bee or wasp sting, a sting from a fire ant can be quite painful. Often it causes itching and a red area around the spot, followed by a pustule that takes several weeks to heal. It may leave a tiny scar. A harvester ant sting can result in a small, purplish spot followed by a blister; pain may persist for up to six hours. Some people become sensitized after being stung repeatedly and become hypersensitive to fire ant and harvester ant stings. They may experience systemic symptoms such as nausea, feelings

Fire ants not only bite, but they also bend the tips of their abdomens around to inflict a sting.

of tightness in the chest, and shock. A field ant bite causes local short-lived pain without any additional symptoms.

It's not the actual bite of these ants that causes the most pain. While a fire or harvester ant is holding on with its jaws (mandibles), it bends the tip of its abdomen around and then stings. Field ants spray formic acid directly into the wound made by the bite. Most multiple ant stings occur when a person sits or stands directly on a nest.

Fire ants can be vicious. They will attack and kill young, unprotected animals such as newborn calves, pigs,

Be cautious when walking in areas where fire ants nest. They will swarm out of their mound to sting any animal or person who steps on the nest.

poultry, and wildlife. Both fire ants and harvester ants can kill a very young animal if it provokes them by walking over one of their colonies or lying down on the nest.

Fire ants have some beneficial aspects—their primary diet is insects and spiders. Studies have shown that they can help control certain crop pests. On the other hand, fire ants eat practically anything, including plants. They sometimes gnaw into and injure roots, stems, buds, or fruit.

The Pest
Fire ants form large colonies composed primarily of worker ants. The workers are reddish brown to black ants about ¹⁄₁₆ to ¹⁄₄ inch long. The colonies build nests of mounded earth 1¹⁄₂ feet or more high and 2 feet wide.

Workers vigorously protect their nest, rushing out in large numbers and swarming over any animal or object that disturbs the mound. If a person walks over the mound, the ants quickly move up his or her legs, stinging exposed skin or stinging through thin clothing.

Several species of fire ant live in the United States; they are similar in appearance, and it takes an entomologist to distinguish them. The most notorious and prolific species—the red imported fire ant—is native to Brazil and appeared in Alabama and Florida sometime before 1940. It now occurs throughout much of the Southeast

and as far west as mid-Texas. Low winter temperatures will prevent the red imported fire ant from invading most of the United States, but this pest could still spread farther north along the Atlantic Coast and to Arizona and the Pacific Coast states.

An average red imported fire ant colony contains 100,000 to 500,000 workers and as many as several hundred winged ants. The ants tend to build their colonies in open, sunny areas, although they may temporarily move in or under a home during periods of rain or drought.

Two species of fire ant, the southern fire ant, which ranges from southern California to South Carolina, and the tropical fire ant, which ranges from Texas to South Carolina, are native to the United States. Both of these species build smaller mounds than those of the red imported fire

Harvester ants keep the area near their nest cleared of vegetation so that the sun will dry and warm the soil above their nurseries.

ant. They become less numerous when the red imported fire ant moves to their area.

Harvester ants are most common in the West; they

remain outdoors, feeding on seeds and storing them underground in their nests. A harvester ant nest looks like a crater in the ground and is surrounded by a large, clear area from 3 to 35 feet in diameter. This area may be strewn with small pebbles and seed husks. Distinct ant trails radiate out from the cleared area in all directions. The worker ants keep the trails and central area bare by girdling the vegetation.

Harvester ant workers are ¹⁄₄ to ¹⁄₂ inch long and red, dark brown, or black. They cannot sting through clothing or tough skin.

Field ants build nests of large mounds of leaves and sticks around rocks or small trees and shrubs. They feed on honeydew and other insects and never nest indoors, although workers may occasionally wander in looking for food. Field ants are common throughout the United States, particularly in mountainous recreation areas. Various species of these medium-sized ants are brown, black, red, or combinations of these colors.

Prevention and Control
Generally, the best way to control these ants is to apply pesticide to the nest entrances (or mound, in the case of fire ants) as well as to an area 4 feet in diameter around the nest. You can drench outdoor

nests of red imported fire ants with insecticides such as Dursban® (chlorpyrifos—Ortho Lawn Insect Spray or

Ortho Fire Ant Control), diazinon (Ortho Diazinon Insect Spray), or acephate (Orthene® Systemic Insect Control or Orthene Fire Ant Killer). If you prefer to use a granule, choose diazinon (Ortho Fire Ant Killer Granules). After applying the granules, be sure to water them in thoroughly.

For successful fire ant control, you must apply the insecticide when the ants are in the top part of the nest rather than when they are deep in the ground. When the soil is very dry, the colony stays deep. Since fire ants don't like extremely wet soil, you can drive them up by thoroughly soaking the ground around the colony. On hot summer days the ants come to the top part of their nests during the early morning and late afternoon, and so these are the best times to treat. During very cold weather they stay down where it is warmer, only coming to the top on a warm, sunny day.

When treating a fire ant mound, be careful to disturb it as little as possible. If you disturb the mound, you increase your chances of being stung, but also any disturbance may cause part of the colony to move deeper and escape the control measures.

For controlling other ants, choose products labeled for outdoor use against ants, such as Dursban® (chlorpyrifos—Ortho-Klor® Indoor & Outdoor Insect Killer) or diazinon (Ortho Diazinon Soil & Turf Insect Control, Ortho Diazinon Insect Spray, or Ortho Diazinon Granules). Use them according to label directions.

First Aid

To treat ant stings, wash the area with soap and water, and apply a disinfectant. Be sure to use disinfectant on any bite that forms a pustule, especially after it breaks. Sensitive people should consult a physician.

ASSASSIN BUGS

The term "assassin bug" is applied to not only true assassin bugs but to the closely related conenose bugs, corsairs, and wheel bugs. Although all of these species can bite, only a few are blood feeders; the rest prey on other insects and are beneficial predators.

Conenose bugs suck blood. The bite is painless while the bug is drawing blood, but later it can cause localized itching and some swelling. Particularly sensitive people may have broader systemic effects, such as nausea, rapid breathing, and increased pulse rate.

Conenose bugs usually come out at night to seek a blood meal and hide in cracks and crevices during the day. They are sometimes called kissing bugs because they may bite around or on a person's lips. They do not feed through clothing, so bites are limited to exposed parts of the body.

Conenose bugs feed mostly on a variety of small mammals. Wood rats are one of their main hosts, and these bugs are especially common in and around their nests. Wood rats build nests in piles of sticks and limbs in forested areas, and in piles of cactus pads or under homes in desert regions.

Conenose bugs are most likely to invade a home in the summer when they disperse from wood rat and other rodent nests. If a nest is next to or under a house, they may get inside by crawling through a crack. Conenose bugs farther from a home are often attracted to outdoor lights at night. The lights confuse their navigational systems, much as they do those of moths. Once the lights have attracted them, some fly in open doors or windows or crawl in the next morning to avoid sunlight.

In Central and South America, conenose bug bites can transmit an organism that

Adequate screening and yellow porch lights will help to keep assassin bugs from flying into the house. These bugs have a painful bite.

causes Chagas' disease. The conenose bugs in the United States do not transmit this disease, however.

Other types of assassin bugs bite strictly to defend themselves. Most bites from these bugs occur when they are accidentally picked up along with vegetation or other debris. They do not transmit any diseases. The bite hurts instantly and continues to hurt for several hours or more, but there are fewer cases of systemic effects than with conenose bugs.

The Pest

A mature conenose bug is $\frac{3}{4}$ to $1\frac{1}{2}$ inches long, with a narrow head and slender nose or beak. Its wings are folded back over the body. An immature conenose bug resembles the adult, except that it is smaller and lacks wings.

The bugs lay eggs during the summer in rodent burrows or other outdoor sites. They take six months to two years to reach adulthood. Conenose bugs rarely fly once they have mated and located a host.

Corsairs, assassin bugs, and wheel bugs are so similar to conenose bugs that it takes an entomologist to distinguish them. Some corsairs have an orange spot on each wing, and some of the assassin bugs glue dirt and debris on themselves for camouflage. Wheel bugs have a very distinctive 8- to 12-toothed "cogwheel" on the thorax, and may be up to $1\frac{1}{2}$ inches long.

All of these bugs have life cycles and habits similar to

those of conenose bugs, except that they prey on other insects and therefore are not associated with animals or their nests.

Prevention and Control

Making your home insect-proof by using tight-fitting screens and sealing up cracks and other openings will help keep conenose bugs out. Remove or control animals, especially wood rats, other rodents, and birds, if they are around or under your home. Treat their nests with an insecticide before you control the animal so that you eliminate the bugs before they disperse in search of a new host. Because the conenose bug is not a common pest, you won't find it listed on any insecticide label. Ask the local cooperative extension service to recommend an insecticide if you need to control bugs in nests or already indoors.

Screen any crawl-space openings to keep animals out, and make any other structural changes necessary to discourage wild animals from nesting under or next to the house. Avoid attracting conenose bugs to the house at night by using few outdoor lights or by substituting yellow light bulbs for white ones, at least during the summer months.

Because corsairs, assassin bugs, and wheel bugs are not common and generally try to avoid people, control is seldom necessary. Prevent bites by not handling the bugs and by wearing leather gloves when working in areas where they are common. If one lands on you, gently brush it away; it may bite if you slap it.

First Aid
To treat bites from any of these bugs, wash the bite area well with soap and water, and then apply an antiseptic. Take a mild analgesic to relieve pain. Antihistamines may relieve systemic effects caused by conenose bugs. Consult a physician in the event of an allergic reaction.

BED BUGS

A bed bug bite is initially painless, and about 20 percent of the people bitten show no subsequent reaction. However, most people develop a small, hard, swollen area where they were bitten, and they may experience some itching. People who are particularly sensitive experience intense itching, and a smaller percentage have severe swelling and a general systemic reaction. Bed bugs are not known to transmit any disease organisms to humans or pets.

Bed bugs bite to obtain a blood meal, taking from 3 to 10 minutes to completely engorge themselves. Early during their feeding, they inject a salivary secretion that makes it easier for them to suck the blood out. This secretion, and not the puncture itself, is responsible for the swelling and itching most people experience.

Bed bug bites can be distinguished from flea bites because they do not have red centers. Flea bites leave a red central area surrounded by a reddish halo.

Bed bug infestations usually occur when the bugs are brought in unknowingly on infested clothing, furniture, or luggage. This is most likely if these items were previously in a place infested with bed

bugs. Hotels, movie theaters, and public transportation terminals are some of the possible sources of bed bugs. They are more often a problem in homes lacking general cleanliness, but even clean homes can become infested.

Because their bites are painless and some people have no skin reaction to them, an infestation may be unnoticed for a time. The first indications may be the dark brown stains of partially digested blood that the pests leave on bed sheets. Homes long infested with bed bugs have a characteristic unpleasant bed bug odor.

Several other bugs—swallow bugs, bat bugs, and poultry bugs—are closely related to bed bugs and can also infest homes. Only an entomologist can tell them apart. Unlike bed bugs, these bugs do not make people their primary hosts. They become pests when their hosts nest in or near a house; when the birds or bats leave, the insects may come indoors.

Swallow bugs live during the summer in swallow nests. They usually leave the nests when the birds migrate in the

fall, although in heavy infestations they may leave several weeks before the birds do. If the nests were attached to a home, the bugs may then come indoors and bite people.

A similar species breeds in chimney swift nests, and another, less common, one, sometimes called the poultry bug, breeds in the nests of a variety of birds, including chickens and owls. Bat bugs feed primarily on bats, but they can sometimes become so numerous in bat-infested attics that they disperse and bite people.

The Pest
Bed bugs are wingless, rusty red, flattened insects about $\frac{3}{16}$ inch long and $\frac{1}{8}$ inch wide. After feeding they become brighter red, plumper, and elongated.

They avoid strong light and are most active at night when they search for a blood meal. During the day, bed bugs prefer to hide in the tufts and folds of the mattress and bedding, in cracks in the bedstead, and in other places close to their food source. Other hiding places include behind baseboards, in fur-

Bed bugs probably first lived with humans in caves; now they infest homes. They are most likely brought in on items from infested public places.

niture, behind pictures and posters, and in any other cracks and crevices in a room. You can identify hiding places that they have used for a long time by the bug's cast skins and black or brown spots of dried excrement.

Bed bugs lay their eggs where they hide. The eggs hatch in about a week in warm weather. The nymphs look like the adults and have similar feeding habits. They molt five times before reaching maturity and live several months to a year. Bed bugs can go for several months without feeding, but they generally feed every few days.

Prevention and Control
Control bed bugs by locating their hiding places and spraying these areas with insecticide. Many ant and roach sprays also control bed bugs; read the label and apply according to label directions. Be especially careful to treat all possible hiding places around and under the bed. Spray the deep seams and folds in mattresses and stuffed furniture, but avoid soaking them. Air out treated bedding and furnishings for at least four hours before use. Do not treat an infant's bed or crib. Re-treat in two weeks or as the label directs if any bed bugs remain. If the home is heavily infested, contact a pest control operator to make the treatments.

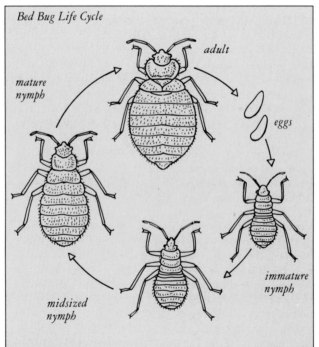

Bed Bug Life Cycle

adult

mature
nymph

eggs

midsized
nymph

immature
nymph

Bed bugs fasten their eggs in crevices with a kind of cement. Nymphs hatch and molt five times, needing a blood meal for each molt. Well-fed adults can live from several months to one year without feeding.

Before using insecticide, launder all bedding materials, thoroughly vacuum the floor, mattress, bed frame, and nearby furniture. Where practical, after spraying seal up cracks or crevices where the bugs may hide. Remove any bird nests or bat roosts in or around the house if bugs are coming from these sources.

First Aid

Washing the bites thoroughly with soap and water is usually the only treatment needed. You can reduce any itching by dabbing the bites with calamine lotion or cortisone cream. In the rare case that systemic or anaphylactic shock occurs, seek professional medical assistance immediately.

BEES

In most people, a honey bee or bumblebee sting hurts instantly, but the pain and any minor local reddening, swelling, or itching disappear within a few hours. Some people are more sensitive. In a moderately sensitive person, the whole limb may swell. If a moderately sensitive person is stung repeatedly or a highly sensitive person is stung just once, he or she may experience anaphylactic shock, which can cause death unless it is treated immediately by a physician.

Most people become less sensitive to bee stings the more often they get stung. However, in about 2 percent of the population, repeated bee stings cause people to become hypersensitive, and each additional sting causes a more severe reaction.

Bees seldom sting while away from their hives unless they are stepped on or otherwise threatened. Most bee stings are inflicted by bees protecting their colonies. This can happen, for instance, when someone prunes shrubs containing a hive or unknowingly steps on a nest of bumblebees in the ground.

Bumblebees are larger and furrier than honey bees. Although they are not aggressive, their sting is painful.

Besides their potential to sting, bees can also be a problem if they nest in wall voids and similar places in a building. Their status as a pest is minimal, however, compared to their beneficial qualities. Bees are important crop pollinators and honey producers. Honey bees alone pollinate 80 percent of the crops pollinated by insects.

The Africanized or Brazilian honey bee—sometimes called the killer bee by the media—is a hybrid of the European honey bee and the African honey bee. These bees are identical to other honey bees in appearance, and their venom is no more potent than that of other honey bees. However, they are aggressive and sting with less provocation and in greater numbers. Africanized bees chase a person or animal farther than other bees and can take half an hour to become peaceful again. Other honey bees take just a few minutes.

A geneticist who was trying to breed better honey producers brought African honey bees from Africa to Brazil. These escaped in 1956 and have been interbreeding with European honey bees and slowly spreading northward into Central America. Africanized bees have not yet established themselves in the United States. Researchers and quarantine experts are attempting to delay or prevent their arrival. Bee investigators disagree over whether these bees actually pose a threat to honey production and crop pollination now carried out by European honey bees.

The Pest

Honey bees are orange and black or brown and were originally brought to North and South America from Europe. A honey bee has a barbed stinger that becomes so firmly anchored in the skin of its victim that the stinger and its accompanying poison sac tear off of the bee. The insect dies a short while later.

Honey bee colonies consist of one reproductive female, called a queen, and as many as 80,000 individuals, which include bee larvae, workers, and a few drones (males). Drones are pushed

Most honey bee stings occur when a person steps on or accidentally grabs a bee. Use caution when walking barefoot or when picking flowers.

out of the hive after the spring swarming season. The queen mates just once in its lifetime. It does this in midair in early spring.

Whenever the colony produces a new queen, either the old queen kills it or one of them leaves with some of the workers to find a new hive site. Swarms may fly a mile or so before settling permanently. They often rest in a tight cluster on a tree branch or other object for several days while scout bees search for a hollow tree or other potential site.

Honey bees collect nectar and pollen from flowers to use as food. Honey, the principal food of the workers, consists of a small amount of pollen mixed with nectar that is "ripened" by bees to reduce excess water. Bee bread, the principal food of the larvae, is similar to honey but contains additional secretions. Larvae that will become queens are fed royal jelly exclusively; this is a substance made of pollen and some honey.

Bumblebees are larger than honey bees, with a robust shape and black and yellow or orange markings. Their colonies are smaller than honey bee hives and are often located in old rodent burrows in the ground or in abandoned rodent or bird nests above ground. Only young, fertilized females survive the winter to establish new colonies the following spring. Bumblebees are less aggressive than honey bees, but their stingers do not have barbs, so they can sting repeatedly. Their sting is similar in intensity to a honey bee sting.

Prevention and Control

Since bees are highly beneficial, it is best not to destroy an unwanted hive. Call a commercial beekeeper to remove it. Beekeepers usually list themselves with various agencies, such as cooperative extension offices and police and fire departments. You may also find them listed in the telephone directory.

A commercial beekeeper can remove a living colony by placing a one-way cone at the hive entrance, with an empty hive next to it. The bees will be unable to get back inside their old hive and will become established in the new one. The cone can be removed after about four weeks, and the bees will then transfer the honey to the new hive. After the bees have moved completely, close all holes and cracks in the old hive site to prevent another swarm from entering.

You may be able to convince a swarm to move on by setting up a garden sprinkler

Although a bee swarm looks threatening, honey bees are perhaps least likely to sting when they are swarming.

so that it rains on the swarm. If you allow a swarm to remain while the scout bees search for a suitable hive site, cover chimneys and any openings in attics and wall voids so that the swarm does not take up permanent residence where you don't want it.

Don't spray an insecticide into a hive located inside your walls. The honey and dead bees will attract scavenger insects that can then become pests. The honey may melt on a hot day and seep inside your home. Call a beekeeper to remove the colony.

You can do several things to reduce the chance of being stung. Never go barefoot on a lawn that has weeds in flower, or on any lawn or other area that has standing water that bees may be drinking. Do not swat at a bee—calmly move away from it or shoo it away with a gentle motion of your hands. Avoid using plants attractive to bees around doorways, screenless windows, paths, pools, patios, and other high-traffic areas. Remove bee nests and swarms from these same areas. Do not leave sweet liquids or garbage exposed outdoors, especially during the late summer when the bee colony is large and there are fewer flowers to provide nectar.

You can discourage bees lapping up water around the edge of a swimming pool or birdbath by providing a separate water source closer to their hive.

First Aid

Immediately after being stung by a honey bee, scrape the stinger out of the skin with a fingernail or a knife. Do not attempt to pull the stinger out with your fingers—this squeezes the poison sac and pushes in more venom. Once you have removed the stinger, wash the site with soap and water. Apply a paste of baking soda or meat tenderizer and water, or wash the site with rubbing alcohol or ammonia.

Moderately sensitive people should keep the stung limb elevated and apply ice packs to it to reduce swelling. You can relieve a mild systemic reaction by taking an antihistamine. Extremely sensitive people should keep an emergency arthropod anaphylaxis kit handy, and should use it and consult a physician immediately if stung by a bee.

BLISTER BEETLES

Blister beetles are aptly named. Simply handling these beetles can cause a mild rash that may turn into fluid-filled blisters 8 to 10 hours after contact. The blisters either subside within a few days or

The margined blister beetle is one of many similar insects that can cause painful blisters if touched.

rupture. If they rupture, the fluid may cause additional blisters or a secondary infection. Ruptured blisters will heal in about a week if they are properly treated.

When disturbed, the beetle exudes a fluid called cantharidin. It does this by holding its breath and increasing its internal body fluid pressure so that tiny breaks occur in its exoskeleton. The irritating fluid exudes from these ruptures, which the insect can repair. If you brush against the insect or pick it up, the irritating fluid can cause blisters.

Adult blister beetles feed on flowers and the foliage of vegetables, trees, shrubs, and vines. They often move in large numbers and occasionally become numerous enough to defoliate plants. (See Ortho's *Controlling Lawn & Garden Insects* for more information.)

The Pest

There are many species of blister beetle; all have necks (pronota) that are narrower than their heads, as well as wing covers, beadlike antennae, and long legs. They range in length from about 1/4 inch to just under 1 inch and are found in a wide variety of colors and color patterns. Immature beetles have long legs when they first hatch from eggs in the soil. As they grow, their bodies become more grublike—fat with short legs.

Blister beetles appear in late spring and summer and have just one generation per year. The immature stages of many species feed on grasshopper eggs and doubtless aid in their control.

Prevention and Control

Avoid handling these beetles if they swarm in your garden. If one lands on you, allow it to walk off, or blow it off without handling it. Because blister beetles are attracted to light at night, use yellow light bulbs for outdoor lighting if the beetles are abundant.

You can handpick the beetles that are in the vegetable garden and throw them into a bucket of soapy water, but be sure to wear gloves and long sleeves for this. You can also spray Sevin® (carbaryl—Ortho Liquid Sevin® or Ortho Sevin® Garden Spray) on listed plants according to label directions.

First Aid

Immediately wash the affected area with soap and water, and apply a mild antiseptic. Protect the blisters with a bandage. If the blisters rupture, blot up the contained fluid immediately and wash and rebandage the area. Consult a physician if the blisters are large, if children have walked barefoot on the beetles, or if the beetles are ingested.

CATERPILLARS, STINGING

Most caterpillars won't harm you if you handle them. A few hairy or spined caterpillars, such as the puss caterpillar, io moth caterpillar, saddleback caterpillar, and buck moth caterpillar, can cause symptoms ranging from mild itching or skin rash to severe burning, numbness, and swelling where they come in contact with your skin. These symptoms begin immediately or within a couple of hours and may continue for several days. The severity of the symptoms depends upon many factors, including the species of caterpillar involved, the sensitivity of the victim, the amount of pressure that was exerted on the caterpillar, and the portion of the body that the hairs contact.

Stinging caterpillars do not have actual stingers; the toxin is contained in the hairs or

The io moth caterpillar has hollow hairs filled with poison. If you brush against the caterpillar, you will receive a painful sting.

short spines along their bodies. When a person brushes against the caterpillar or picks it up, the tips of the hairs break off in the skin, allowing toxin to flow out of the hollow center of each hair and into the skin.

These caterpillars feed on tree and shrub foliage, and you may encounter them in your yard. They are usually not numerous but, for reasons unknown, large outbreaks sometimes occur.

Be cautious around any hairy or spiny caterpillar, because many other kinds not mentioned here can give a skin rash. Other species usually cause milder reactions than those listed here.

The Pest

The puss caterpillar is completely covered with flattened hairs and is yellow, gray, or reddish brown. It is about 1 inch long when mature. Among the dense hairs are yellowish spines with black tips. The pale green io moth caterpillar has a red or maroon stripe edged with white down each side of its body. It is relatively hairless with scattered tufts of spines and may reach 3 inches long. The buck moth caterpillar is similar to the io moth caterpillar but is purplish black and red. The 1-inch-long saddleback caterpillar is brown with a green and white mark resembling a saddle on its back. The body is hairless but has a large number of spines.

These caterpillars have one or two generations per year.

Depending on the species, they can be found from late spring to fall.

Prevention and Control

You can control caterpillars by spraying the infested plants with an insecticide containing Orthene® (acephate), Sevin® (carbaryl), or *Bacillus thuringiensis* (BT). Be sure that any plants you treat are listed on the product label.

Avoid areas infested with stinging caterpillars, and warn children not to handle them. Wear gloves, a long-sleeved shirt, and long pants when working in an infested area.

First Aid

Immediately after touching the caterpillar, apply tape to the skin and pull it off to

The distinctive markings of the saddleback caterpillar make it easy to identify. If stung, use tape to remove spines from skin.

remove any spines. Do this several times, using fresh tape each time. Wash the affected skin with soap and water, and then apply a paste of baking soda and water. Ice may also help. Consult a physician in the event of severe symptoms.

CENTIPEDES

Some centipedes can inflict a bite similar in discomfort to a bee or wasp sting. The venom is never lethal. There may be some local redness and swelling at the bite site, but this subsides in several hours. Centipede bites generally happen outdoors, where most centipedes live. The one species that can live and breed indoors, the house centipede, rarely bites.

All centipedes prey on other insects and are primarily beneficial. They quickly subdue their prey by injecting venom through two claws located near their mouths. Centipedes are active at night, hiding during the day in damp, dark places such as under stones, leaf mulch, and logs. Indoors, house centipedes may occur throughout the home, but they are most common in damp areas of basements and bathrooms and places where they can find insects.

The Pest

Centipedes, or hundred-legged worms, as they are sometimes called, are elongated, flattened creatures with one pair of legs per body segment. Most adult centipedes are longer than 1 inch. They look like millipedes, but millipedes have two pairs of legs per body segment, coil up when disturbed, and move very slowly compared to centipedes. (For more information on millipedes, see page 45.)

The house centipede has longer legs and antennae than other centipedes. Its antennae and rear set of legs can be more than twice the length of the body. Mature house centipedes are up to 1½ inches long. They run quickly and make sudden stops.

Adult centipedes spend the winter in moist, secluded places. They lay their eggs in damp soil in the spring and summer months.

House centipedes dart about quickly, then stop suddenly. They thrive in damp parts of a house.

Prevention and Control

Centipedes seldom need to be controlled unless they become abundant indoors. Indoors, treat along baseboard cracks and other hiding places with a combination of tetramethrin and sumithrin (Ortho Household Insect Killer Formula II or Ortho Home & Garden Insect Killer Formula II), resmethrin, pyrethrins, bendiocarb, or Baygon® (propoxur). If centipedes are coming from outdoors, apply Sevin® (carbaryl), diazinon, Baygon® (propoxur—Ortho Pest-B-Gon® Roach Bait), or Dursban® (chlorpyrifos) in a band around the outside of the foundation or directly to moist hiding places and to points of entry, such as crawlspace vents and cracks in concrete slabs.

You can reduce the number of house centipedes indoors by reducing other household pests that they are feeding on. Eliminating excess moisture from leaks in plumbing, air conditioning units, and similar sources also helps control them. Outdoors, remove organic matter around the foundation so that they will not have a moist place to hide and breed.

Always shake out or invert shoes, clothes, and sleeping bags before using them if you are in an area where centipedes are common.

First Aid

Wash the bite area with soap and water, and apply an antiseptic.

FLEAS

Fleas give blood-sucking bites that can make dogs and cats miserable with constant itching and scratching. Some pets also suffer from flea allergy. The hallmark of such an allergy is an open sore at the base of the pet's tail. Red skin and hair loss at the tail base are also common.

Adult fleas spend about 80 percent of their time on an infested animal. They can subsist outdoors by feeding on wild animals and indoors by feeding on pets or people. Once fleas get into a house, they can become a serious nuisance that is difficult to control.

Fleas have definite preferences for the kind of animal from which they suck blood. When their favorite hosts are absent, however, the fleas look for substitutes. Cat and dog fleas prefer these and other animals but also readily feed on humans, especially if the fleas can't find their usual host. People are bitten by fleas most often after a pet dies or is absent from the house for several weeks. Likewise, when a rat or wild animal nesting in or under a home dies or leaves, fleas can come indoors from the dead animal or its nest and proceed to bite people.

Most people get fleabites around their ankles and lower legs, but if you sit on an infested chair or lie on the ground, you may be bitten anywhere on exposed skin.

People vary greatly in their reaction to fleabites. Many fortunate people don't even notice the bites, but most people experience itching and develop a small red spot where the flea's mouthparts have penetrated the skin. A characteristic red halo surrounds the spot, and there is little or no swelling. Particularly sensitive people suffer intense itching for up to a week and sometimes come down with a generalized rash. People often become immune to fleabites with repeated biting, but a very small percentage of people become more sensitive.

Besides the uncomfortable bites, on rare occasions fleas can transmit disease organisms that cause bubonic plague and murine typhus. They can also carry tape-

Eyeless and legless flea larvae live on organic debris in pet bedding. They must be eliminated along with the adult fleas.

worms and transmit them to people or pets who accidentally ingest the fleas.

The Pest

Cat fleas are the most common fleas around homes. Contrary to what the name implies, they infest both cats and dogs equally. Dog fleas also infest both cats and dogs but are more rare. Sticktight fleas and human fleas are sometimes problems on pets in more rural areas. Various species of rat flea occur on rats and other rodents. People sometimes say they have sand fleas, but this name doesn't refer to any particular species. Many fleas, including cat fleas, do very well in sandboxes and other sandy areas. The "sand fleas" common at beaches are not fleas at all but harmless amphipods.

Adult fleas are about $\frac{1}{10}$ inch long, black or brown, wingless, and flattened from side to side so that they can run quickly in between hairs. They have strong hind legs that enable them to jump vertically 4 inches or more and attach themselves to clothing or skin.

Female fleas lay from four to eight eggs after each blood meal. They drop some of their eggs in the host's sleeping areas, but they lay most on the host. These eggs are not firmly attached to the host and eventually fall off wherever the host happens to be. The small, white or grayish, legless larvae hatch in 2 to 12 days. They feed on various kinds of organic debris, such as lint, hair, and especially the dry bits of partly digested blood that adult fleas excrete.

Common areas indoors for the larvae to feed include pet sleeping areas, grooves in the floor, under furniture, and along baseboards and rug edges. Outdoors they occur in areas frequented by infested animals. They may be in pet sleeping areas, in crawl spaces, under shrubs, along pet walkways, and in similar moist areas protected from hot, dry conditions. Flea larvae do not survive in areas that receive frequent, heavy watering, which can drown them.

The larvae spin tiny cocoons with their own silk and bits of fabric, sand, or other debris. These added materials help camouflage the cocoon. Adult fleas can emerge from the cocoon as soon as one week after pupation but they may lie dormant in cocoons for many months. The adults often remain in their cocoons until a host is present. It takes a flea about two weeks to go from egg to adult at 95° F, but the same process takes almost five months at 55° F.

Prevention and Control

To control fleas successfully, you must treat your pet and all indoor and outdoor areas likely to harbor fleas at the same time. Expect a much quicker reinfestation if you control fleas in only one area and allow them to remain elsewhere.

Control fleas on your pets by dusting, spraying, or shampooing them with pesticide. Choose a product containing Sevin® (carbaryl—Ortho Sevin® Garden Dust), tetramethrin plus sumithrin (Ortho Flea-B-Gon® Flea Killer Formula II), pyrethrins (Ortho Pet Flea & Tick Spray Formula II), malathion, d-limonene (an effective citrus oil derivative), or one of the other flea products labeled for use directly on pets. Follow label directions and cautions exactly, and never treat a pet

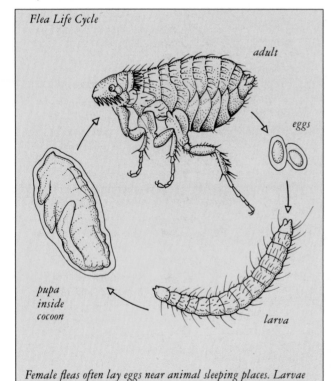

Flea Life Cycle

adult

eggs

pupa
inside
cocoon

larva

Female fleas often lay eggs near animal sleeping places. Larvae feed on debris, then spin cocoons in which they pupate into adults. Both sexes feed on blood; females need it to produce eggs.

with a product that does not specifically state that it can be used on animals. Re-treat the pet when it seems to be carrying fleas again.

If your pet is in frequent contact with other infested animals, you may need to treat it as often as every week or two, but never apply the insecticide more often than the label recommends.

Many veterinarians will treat pets for fleas. In some cases they may prescribe adding Proban® (cythioate) to the pet's food. This systemic insecticide travels through the

This greatly magnified flea, in typical head-down biting position, is feeding on a dog.

pet's blood and kills fleas when they bite. It will not prevent flea allergies, however, since the flea must bite to be controlled.

A flea collar is useful as a supplement to treating the pet. Flea collars are only partially effective, especially on large pets where the collar is far away from the animal's tail end. Do not allow pets to chew on the collars. Some animals are sensitive to flea collars; check to see if your pet develops a skin rash when wearing one. Follow label directions.

Your pet's regular sleeping quarters are usually infested by more adult fleas and flea larvae than other areas. Wash the bedding material in hot, soapy water, and spray or dust the sleeping area with tetramethrin and sumithrin (Ortho Flea-B-Gon® Flea Killer Formula II), Dursban® (chlorpyrifos—Ortho Home Pest Insect Control or Ortho Flea-B-Gon® Flea & Tick

Killer), Sevin® (carbaryl—Ortho Sevin® Garden Dust), malathion, pyrethrins, or methoprene.

Vacuum all floors, carpeting, and furniture in rooms where the pet goes. Dispose of the vacuum bag immediately. Use one of the above insecticides according to label directions, or choose a total-release aerosol, such as Ortho Hi-Power Indoor Insect Fogger. A total-release aerosol is one way to treat a widespread infestation, but be sure you use the proper number of canisters to treat each room where you suspect there are fleas. (See page 17 for more information on total-release aerosols.)

An excellent new flea insecticide is methoprene, an insect growth regulator that prevents the larvae from pupating into adult fleas. Because only the adults bite and reproduce, this compound is highly effective in controlling fleas. It is usually combined with a traditional insecticide to kill existing adult fleas. Methoprene is included in some total-release aerosols and flea sprays.

Control fleas outdoors by applying insecticide to crawl spaces if the pet has access to them, around pet sleeping areas, along pet pathways, and similar areas. Fleas can be anywhere, but most of them will be in areas that your pet (or a wild animal) frequents. Fleas are especially likely to infest moist areas that are never flooded by rains or heavy watering and are protected from direct sun. Use Sevin® (carbaryl—Ortho Liquid Sevin®), diazinon (Ortho Diazinon Insect Spray), malathion (Ortho Malathion 50 Insect Spray), or Dursban® (chlorpyrifos—Ortho Home Pest Insect Control or Ortho Outdoor Ant, Flea & Cricket Spray). Do not allow your pet to reenter the area until the spray has dried.

Good sanitation can greatly reduce flea popula-

Specially made flea combs remove fleas and eggs from a pet's fur. Kill fleas by cleaning the comb in water containing detergent.

tions and prevent their build-up. Vacuum carpets and behind the cushions of upholstered furniture where pets climb, and any other areas that may contain adults, eggs, larvae, or debris on which the larvae can feed. After every vacuuming of an infested area, empty the vacuum bag of its contents, because immature fleas continue their development in the debris in cleaner bags.

To prevent reinfestations, screen all crawl-space vents to prevent access by pets, rodents, and stray animals. If practical, do not allow infested animals to enter your home or yard, and keep your pets away from infested areas. Clean all pet bedding materials and the area surrounding its sleeping area every week or two. Vacuum rugs regularly, and mop or sweep other areas. Treat your pets on a regular basis with a flea shampoo or dust according to label directions, and/or use a flea collar.

Studies show that feeding pets a vitamin B-1 supplement reduces flea feeding on some animals but not on others. Removing fleas by combing your pet with a special flea comb and dropping the captured fleas into soapy water may help reduce fleabites and flea populations, especially if you do it thoroughly and on a daily basis.

Keep in mind that successful control means eliminating fleas in three places: on the pet; in the house, especially in the pet's sleeping quarters; and outdoors in moist, shady

areas frequented by the pet. If you miss any of these spots, fleas will probably not be controlled. If a person in the household is particularly sensitive to fleabites, or if your own control methods have not succeeded, consider hiring a professional pest control operator to control fleas.

First Aid

For fleabites on people, relieve any itching by applying a cooling preparation, such as menthol, camphor, calamine lotion, carbolated petroleum jelly, or cortisone cream. See a physician if an infection or any unusual symptoms occur. Antigens are available for people who have severe reactions. These treatments are so effective that the person becomes unaware of fleabites.

A veterinarian can treat dogs and cats that suffer from flea allergy.

FLIES, MIDGES, AND GNATS

Horseflies and deerflies are the largest biting flies, and they have the most painful bites. These insects actually cut their victims' skin and then lap up blood from the open wound. The bites often continue to bleed after the fly has left. The pain is immediate but usually short-lived. Unusually sensitive people may experience high fever and headache.

Stable flies suck blood, usually on the lower part of the body around the ankles. The bites are very painful and

Above: Deerflies are similar in habit to horseflies, but are smaller and have striped wings.
Below: Although female horseflies mainly bite animals, they sometimes bite people. Most species are active during the warmest part of the day.

may bleed profusely, but they rarely cause any lasting discomfort or systemic symptoms.

Blackflies and biting midges such as sand flies pierce the skin of their victims and suck up blood. Their bites are painful, itchy, and accompanied by local swelling. A small red spot develops at each bite site, followed by an itchy weal. Headache, fever, and nausea can develop in sensitive people.

On many people, sand fly bites develop into small, clear blisters on the second day. The blisters disappear a day or two later.

Eye gnats neither suck blood nor cut into flesh to cause bleeding. They sponge up body fluids from around eyes and open wounds. When they suck up fluids from the eyeball, they produce many tiny scratches. If bacteria enter these scratches, a condition known as pinkeye may develop.

These flies are annoying, and their bites can cause severe discomfort. If horseflies,

deerflies, blackflies, or stable flies are numerous enough, their bites can kill livestock and wild animals, which are unable to protect themselves from the flies. In some cases the flies transmit diseases that kill the animal. In other cases the animals die from a combination of a massive loss of blood from the fly bites and the discomfort caused by the saliva that the flies inject to make the blood easier to suck up.

The Pests
These insects are all flies and thus are characterized by having a single pair of wings and undergoing four phases of development: egg, larva or maggot, pupa, and adult.

Horseflies look like huge house flies and are the largest biting flies, measuring from $1/2$ inch to $1\frac{1}{8}$ inches long. Deerflies are $3/8$ to $5/8$ inch long. You can distinguish them from horseflies by their smaller size and the dark markings on their wings. They

have similar life cycles. Both sexes feed on plant juices and nectar. The females also suck blood from a wide variety of animals, preferring horses, cattle, and deer. If these animals are absent, horseflies will attack and bite other animals and people.

Horseflies lay eggs on low-growing plants or other objects near the edges of ponds, lakes, or calm streams. The larvae feed on other insects or decaying vegetation and pupate in damp plant debris. Horseflies are slow to develop, and many require a year or more to complete their life cycle. Deerflies develop more rapidly and have two or more generations per year.

Horseflies and deerflies usually bite during warm parts of the day, although a few species have peaks at dawn and dusk. None bite at night. They find their host by following large, dark, moving objects.

Stable flies are about the same size and color as house flies. Both sexes feed on the blood of humans and a wide variety of animals. Females lay their eggs in piles of soggy hay and grass clippings and other accumulations of damp, decaying organic material. The larvae feed on these same materials. These flies can complete a generation in one month or less.

Stable flies can be a problem indoors, especially in the autumn and during rainy weather. They are common along shorelines, where they breed in piles of rotting seaweed, and in areas where a large number of domestic animals provides the adults with a constant food supply. They breed in manure that is mixed with straw or hay.

Blackflies, also called buffalo gnats and turkey gnats, are black or gray humpbacked flies $1/16$ to $1/4$ inch long. The males feed entirely on nectar from flowers, but the females also suck blood from many mammals, including people.

They lay their eggs in rapidly flowing water such as streams and irrigation canals. The larvae eat bacteria and small particles of organic matter. The adults emerge during late spring and early summer. Depending on the species, there may be one to five generations per year.

Blackflies usually bite exposed areas of the body, often at the edges of snug-fitting clothing; just underneath the rim of a hat is a favorite place. They bite during the day, and since they often are found in large swarms they can make it impossible to work outdoors. They sometimes attack children more than adults. These flies are most abundant near their breeding sources but can fly and be blown by wind for 10 miles or more. They sometimes enter homes accidentally and crawl about on windowpanes until they die.

Sand flies, also known as no-see-ums, punkies, black gnats, and biting midges, are

Both male and female stable flies feed on human blood.

only $1/25$ to $1/8$ inch long and are usually so small that you feel their bite but have to look very closely before you see them. The males do not feed on anything; females suck blood from humans and other mammals as well as birds. Sand flies bite exposed areas of the body and also crawl underneath loose clothes and bite where the material fits tightly. They bite any time from late afternoon to dusk and early evening, depending on the species.

It may be impossible to work out of doors when swarms of black-flies are active. The females bite humans and other mammals.

Sand flies deposit their eggs in damp soil around marshes and streams, in rotted pockets of tree wood, and in decaying plants. The larvae feed on decaying organic matter and small insect larvae. Adults begin emerging in late spring. Sand flies complete a generation in one to two years. They do not fly far from where they lived as larvae. Sometimes you can avoid their bites simply by moving a few yards away.

Eye gnats are about 1/16 inch long. They lay eggs in moist soil, generally within a few hours after it is plowed. The larvae feed on decaying

organic matter. Eye gnats can complete a generation in about three weeks. The adult gnats are most numerous near their larval breeding grounds, but some may be blown as far as five miles away.

Prevention and Control

To prevent any of these pests from getting into your home, or to control ones already inside, use the same techniques recommended for house flies (see pages 43 and 44).

To keep populations down, try to eliminate breeding sites. For horseflies, spread and dry out piles of debris that can be suitable for the larvae. Horseflies and deerflies are difficult to control; draining nearby ponds and ditches to kill the larvae can help, but the adults can fly long distances. Reducing moisture in likely nearby breeding places can keep numbers of sand flies low. As is the case with mosquitoes, preventing blackfly populations from developing usually requires the large-scale efforts of a government agency.

To control any of these flies on a temporary basis, spray outdoor areas around

Sand flies are tiny; many species bite just as it is getting dark. Often they bite before the victim notices them—hence the nick-name "no-see-ums."

your home with insecticide. Choose tetramethrin (Ortho Home & Garden Insect Killer Formula II), resmethrin (Ortho Outdoor Insect Fogger), or pyrethrins to control listed pests according to label directions. Spraying will provide only temporary relief because most of these pests can fly in from a long distance. A single spraying may control sand flies for the entire year because they do not fly far.

When you are outdoors in areas infested with these biting pests, apply an insect repellent to exposed areas of your skin, wear a long-sleeved shirt and long pants, and tuck your pant legs inside your boots. Use a head net in severe infestations of blackflies, or wear a hard hat with a thin film of oil smeared on the outer surface. It is important to wear two pairs of socks to prevent stable fly bites around your ankles.

Because blackflies are active at specific times of the day, learn when the local species are active and avoid exposure during these periods. One study found that dark blue cloth attracted the most blackflies and white cloth attracted the fewest.

First Aid

Wash fly bites with soap and water and apply an antiseptic. Oral antihistamines can be taken to relieve any mild systemic symptoms. Consult a physician if unusual symptoms or an infection develops.

LICE

Although they are much less common today than even half a century ago, lice still occur, even in the cleanest of families. Three types of these parasitic insects infest people—head lice, body lice, and pubic lice. Anyone can get head lice, even a meticulously clean person. Head lice are easily spread between schoolchildren by close, daily contact and by the sharing of combs, scarves, headgear, and similar items that contain lice or hairs with nits (louse eggs).

Body lice can transmit disease organisms that cause louse-borne typhus, relapsing fever, and trench fever. These diseases are mostly of historical interest, however, and occur only rarely in the United States today.

A person with lice can suffer intense itching and in some cases may develop a rash resembling German measles and experience a general tired feeling. People with a long-term infestation may become sensitized and suffer even more acute symptoms, or they may develop hard, pigmented areas of skin, a condition known as vagabond's disease. If a person scratches and abrades the skin, a secondary infection can ensue. Characteristic blue spots often develop with a crab louse infestation.

The Pest

The three kinds of lice that affect people, the head, body, and crab or pubic louse, are insects that suck blood exclusively from humans. These dirty-white to grayish black, wingless creatures are about 1/8 inch long. They have a strong claw on each leg, which allows them to cling to hair. The eggs, often called nits, are oval and grayish white.

Lice complete a generation in about three weeks. Adults can live about 30 days, but all lice die within a few days if they are away from their host and are unable to get their blood meals.

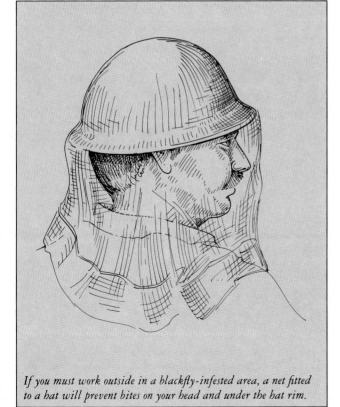

If you must work outside in a blackfly-infested area, a net fitted to a hat will prevent bites on your head and under the hat rim.

Head lice almost always occur on the head, especially above the ears and on the back of the head. They attach their eggs (nits) to the bases of hairs. Body lice spend most of their time in clothing rather than on the body. They periodically crawl onto the skin to feed and then return to clothing to lay eggs.

Body lice feed primarily where skin is soft and folded, such as at joints. They deposit most of their eggs on clothing fibers, especially in the seams next to the body, and only occasionally glue them to coarse body hairs.

People who continuously wear the same clothes, even while sleeping, are most likely to have problems with these lice. Body lice seem to prefer

Insecticidal shampoos, correctly used, have replaced head shaving as a control for head lice. Here head louse eggs cling to human hair.

wool clothing. Infestations tend to reach a peak in winter and are very low in the summer months. These lice may be transmitted by close contact in crowded living quarters and by the sharing of clothing.

Crab lice occur on coarse hairs of the body, primarily in the pubic region but also in armpits and on eyelashes, eyebrows, and mustaches. They attach their eggs to these hairs. Crab lice are spread primarily through sexual contact but can also be transmitted on clothing in crowded locker rooms, in bedding, on towels and similar articles, and, rarely, on toilet seats.

Prevention and Control

To effectively control lice, you must eliminate the lice and nits on both clothes and the head or body. Kill lice on all clothes, towels, sheets, blankets, and similar items that might be infested. You can do this by either dry-cleaning the items or thoroughly washing them in hot, soapy water. Also clean combs, brushes, hats, and similar items that may be infested. Vacuum upholstered furniture.

An alternative to cleaning is to isolate the items until the lice die of starvation. Keeping the items away from people for 30 days kills all nymphs and adults and gives the nits enough time to hatch into nymphs and die for lack of a host. If there are no nits, 10 days of isolation for head and body lice and 2 days for crab lice is sufficient to kill them. Unlike body lice, head and crab lice attach their nits to human hairs, so to remove nits you must remove hairs from an item by handpicking or thorough vacuuming if 30 days is too long to isolate an item.

Washing with ordinary soaps and shampoos will not control lice. A physician or pharmacist can recommend an insecticidal shampoo for head and crab lice or a lotion for body lice. Most of these contain pyrethrins. Avoid applying these preparations over a prolonged period, or they may have undesirable side effects. Follow label directions exactly. Never use a household insecticide to control lice or any other insect on your body.

Insecticidal cleansers are not suitable for lice on eyelids and eyebrows. Remove lice and their eggs in these areas with a pair of forceps, or ask your physician for a milder ointment that won't affect the delicate eye tissues.

In the past, people controlled head lice by shaving the head or having another person carefully handpick the

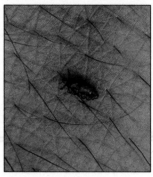

Anyone can get head lice, since lice are easily spread by sharing combs, scarves, or hats.

lice and nits on a daily basis for two weeks. These methods are used less commonly today because of the availability of effective insecticidal shampoos. You may be able to control an infestation of body lice by bathing daily and then dressing in a clean set of clothes each time.

To avoid reinfestation, treat all family members who might have lice, and clean infested items all on the same day. To help keep from acquiring lice, avoid close contact with infested people, their clothes, and their bedding materials.

First Aid

Symptoms will quickly end when you control the lice. Consult a physician if severe or unusual symptoms develop.

MITES AND CHIGGERS

Most mites—very small relatives of spiders—feed on plants, decaying plant material, or insects. Some species, such as chiggers, scabies or itch mites, straw itch mites, and rodent and bird mites, suck blood from people, animals, or both.

Chigger mites cause intense itching beginning 3 to 12 hours after they have attached themselves. The itching may persist for as long as two weeks. The scabies or itch mite causes a rash and intense itching, beginning about a month after the infestation

started. Many other mites affect humans or pets in a similar manner, with the time it takes to develop an itch or rash usually several hours to several days. Some of these mites can cause mange, a thinning or total loss of hair in infested areas of the body.

The Pest

The mite life cycle generally consists of an egg stage, several nymph stages, and an adult stage. The entire life cycle for most mites takes only two to three weeks under favorable conditions. Many of these mites are microscopic or nearly so, and you thus probably won't see one with your unaided eye. A specialist is needed to distinguish between the different species.

Chigger mites are by far the most common mite people encounter. They live in weedy areas, sunny fields, areas where there is lots of lush

An electron micrograph reveals a house dust mite crawling through particles of house dust. Sensitive people react to them with asthma symptoms.

undergrowth, and places where the ground is undisturbed. They are especially prevalent in early summer throughout the country, but particularly in the South.

Chiggers run around on a person's legs for several hours before they begin to feed. Only the young nymphs (larvae) feed on people and animals; the older nymphs and the adults feed on small insects and insect eggs. Chigger bites cause welts and hard, raised bumps, particularly on parts of the body where clothing is binding or where body parts touch one another, such as belt and sock lines, armpits, the backs of knees, and under cuffs and collars.

Contrary to popular belief, chiggers do not burrow into the skin. They attach themselves with special claws located around their mouths. If you look closely, you can usually find a red spot, the center of which may contain a tiny red dot that is the chigger. If undisturbed, chiggers will feed for about three days and then drop off. The intense itching is caused by the saliva that chiggers secrete to keep the host's blood from clotting while they feed.

Scabies or itch mites cause a skin rash called scabies. The rash can be so severe that a person may not be able to sleep. People usually become infested with this mite by living in crowded quarters, through sexual contact, or by sleeping in the same bed as an infested person. These mites actually burrow into the skin of their hosts. Hands and wrists are infested more than other portions of the body.

Certain varieties of scabies mite commonly affect dogs and cats and may then infest people. These mites are especially likely to attack humans if an infested pet has intimate contact with people, sleeping on its owner's bed, for example.

Rodent and bird mites feed primarily on rats, mice, other animals, and birds. On people, they cause painful red spots that continue for two or three days. They are most likely to be a problem when their other hosts are living in, under, or next to a home. The mites drop off animals as they walk around, or they crawl away from the animals' nests, especially when they die. As with many other mites, these may affect certain people in a household and not others.

Straw itch mites cause rashes almost exclusively on clothed portions of the body. They feed primarily on insects, especially ones that infest stored grain, and are much more common if their hosts are present. Bites from these mites are usually associated with sleeping on straw mattresses, harvesting grain,

or otherwise coming in close contact with this type of material. The mites feed on flesh, not blood, and remain attached to human skin for only a short time.

All houses contain house dust mites. These mites feed on human skin debris commonly found on mattresses and in house dust on floors. The mites also live in rugs, upholstered furniture, and other protected places. House dust mites produce asthma symptoms in sensitive people.

Prevention and Control

To determine whether chiggers are present in a lawn or weedy area, place a piece of black cardboard perpendicular to the ground. The tiny yellow or pink nymphs will crawl rapidly over the cardboard and accumulate on the upper edge. They can also easily be seen on black, polished shoes.

To prevent chigger bites, try to avoid known chigger-infested areas. If you must go into them, wear clothing that fits snugly at the collar, wrists, and ankles but is otherwise loose, and apply a repellent, such as diethyltoluamide (deet). Treat exposed parts of your body, including your neck, and your clothing near any openings. Keep lawns and weedy areas around the house well mowed.

Control chiggers in lawns and weedy areas by spraying with an insecticide such as diazinon (Ortho Diazinon Insect Spray) or by applying insecticide granules containing diazinon (Ortho Diazinon Soil & Turf Insect Control).

To control scabies mites, treat all family members, even if they show no symptoms, with the proper medication at the same time. (See First Aid.) Wash all infested bedding and clothing. Mites cannot survive on clothing for more than three days without a host. Avoid intimate contact with infested persons and with their clothes and bedding to keep from becoming infested.

You can avoid rodent and bird mites by controlling rodents and birds in, on, and under your home. Remove bird nests. (See "Rodents and Other Animal Pests.") Treat nests and animal sleeping places for mites before you control the animals or remove the nests. Contact a pest control operator to control the mites, or ask the local cooperative extension service for advice on which insecticide to use in your particular situation.

Prevent bites from straw itch mites by avoiding infested areas. Remove the source of infestation, such as a pile of straw or bale of hay, or contact a pest control operator to control the mites.

House dust mites exist in all homes, but frequent vacuuming may help to control them. Contact a pest control operator if allergies are severe and eradication is necessary.

First Aid

Remove chiggers from the skin as soon as possible after walking through a chigger-infested area by taking a hot, soapy bath. This will kill or

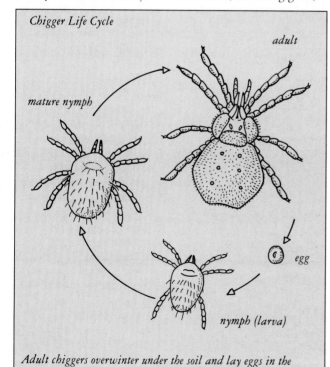

Chigger Life Cycle

adult

mature nymph

egg

nymph (larva)

Adult chiggers overwinter under the soil and lay eggs in the spring. The youngest nymphs, which are known as larvae, have six legs and feed on blood, dropping off when sated. Older nymphs feed on insects and insect eggs, then molt into adults.

remove the unattached larvae and most of the attached chiggers. Then apply an antiseptic, calamine lotion, ammonia, rubbing alcohol, or iodine to any welts to kill any remaining attached chiggers and to prevent infection. Consult a pharmacist for other compounds to relieve the itching. An antihistamine may be useful if you have been bitten by a large number of mites. Consult your physician.

You may need to have a dermatologist diagnose and treat scabies, straw itch, rodent, and bird mites. A physician can prescribe a cream to control the mites. Some itching will occur for up to two weeks after mites are killed. Consult a veterinarian or pet store for a medicine to use on an infested pet.

For treating house dust mite allergies, consult your family doctor or an allergist.

MOSQUITOES

Mosquito bites typically itch and result in a small, red swelling at the bite site. People bitten frequently by mosquitoes usually become more immune to the bites, but a small percentage of people become even more sensitive. Sensitive people may have severe allergic reactions.

Mosquitoes can transmit malaria, encephalitis, yellow fever, dengue fever, and many other diseases that affect people or animals throughout the world. Currently, malaria is absent from the United States, although epidemics have occurred here in the past. Several mosquito species that have the ability to transmit malaria exist in the United States, and the disease could be reintroduced to these mosquitoes at any time by infected travelers returning from countries in which malaria still occurs. Encephalitis is the only disease currently transmitted by mosquitoes in the United States. The disease has a high mortality rate in people, horses, and birds.

Some people are more attractive to mosquitoes than others. There are many reasons for this. Carbon dioxide, perspiration, and heat are strong mosquito attractants. Anyone who produces larger quantities of these than other people do is more likely to be bitten. For instance, someone perspiring from working in the garden will attract more mosquitoes than will a person lounging in a hammock nearby. Many species of mosquito are more attracted to dark-colored than light-colored clothing.

Certain species of mosquito bite men more often than women. Others prefer women to men. Mosquitoes are highly attracted to estrogen and certain amino acids, and blood levels of these can vary from person to person. For instance, women are more attractive to mosquitoes during ovulation because of high estrogen levels.

The Pest
Mosquitoes are small, delicate flies that are less than $1/4$ inch long, with wings covered in tiny scales. Both the males and the females feed on

flower nectar and other sweet substances, but females must also suck blood from animals before they can produce eggs.

Mosquitoes lay their eggs primarily in standing bodies of fresh, stagnant, or salty water. The larvae, or "wigglers," usually remain at the water surface, breathing through an air tube. They eat bacteria, algae, and small bits of organic matter. Larvae develop into pupae, which display a characteristic tumbling movement, rolling head over tail, when disturbed. The adults emerge after two or three days and fly off. A complete life cycle takes 10 days to several weeks for most species, but some species take much longer. Most mosquito species complete a number of generations each year. The females live two weeks to several months.

Many species of mosquito spend their lives within a few hundred yards of where they emerged from pupae, but some fly many miles to search for a blood meal.

The swelling and itching experienced when a mosquito bites is caused by the saliva that the insect injects. This

Mosquitoes numb skin with an anaesthetic as they bite, but later, irritants in their saliva make the bite itch.

saliva contains various substances that help the blood flow more freely, and it also contains an anesthetic to reduce pain while the mosquito is feeding. Mosquitoes whose salivary glands have been cut experimentally cause much more painful bites, but there is no itching or swelling afterward.

Not all mosquitoes feed on humans; some feed solely on reptiles or birds or prefer the blood of other mammals to that of humans. Of the species that suck blood from people, some feed outdoors exclusively while others frequently enter homes to feed. Some species feed only at night; others feed during specific periods of the day.

Prevention and Control
Control adult mosquitoes indoors by spraying or fogging rooms with pyrethrins (Ortho Hi-Power Indoor Insect Fogger), tetramethrin and sumithrin (Ortho Home & Garden Insect Killer Formula II), resmethrin, and d-trans allethrin (Ortho Professional Strength Flying & Crawling Insect Killer) according to label directions.

Outdoors, kill mosquitoes on lawns and in dense shrubbery by spraying with Dursban® (chlorpyrifos—Ortho Lawn Insect Spray), malathion (Ortho Malathion 50 Insect Spray), resmethrin (Ortho Outdoor Insect Fogger), Sevin® (carbaryl), or naled. Control mosquitoes resting on the outside walls of

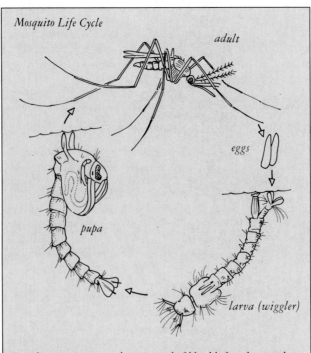
Mosquito Life Cycle

adult

eggs

pupa

larva (wiggler)

Female mosquitoes must have a meal of blood before they can lay eggs. They lay eggs on the surface of the water or near its edge. Larvae, called wigglers, feed on microscopic water life and then develop into pupae. A life cycle can take as few as 10 days.

the house, especially near doorways, to decrease the number that enter when you open the door. Spray with tetramethrin and sumithrin (Ortho Home & Garden Insect Killer Formula II) or pyrethrins. For longer-lasting control on outdoor surfaces of the house, spray with Dursban® (chlorpyrifos—Ortho Home Pest Insect Control), diazinon, malathion, or Baygon® (propoxur).

Be sure that the doors and windows have tight-fitting screens to help prevent mosquitoes from flying indoors. Repair any torn screens.

When you are outdoors in mosquito-infested areas, apply a mosquito repellent containing diethyltoluamide (deet) to any exposed areas of your skin and to any thin clothing through which mosquitoes can bite. You can protect your head by spreading mosquito netting over a hat and attaching the net loosely to your collar.

To prevent large populations of mosquitoes from developing, you must control the larvae. One of the simplest and most effective ways to

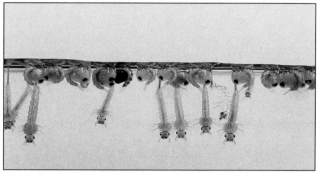

In the event of a mosquito outbreak, carefully examine the yard and even the basement for standing water in which mosquito larvae and pupae may be developing.

control mosquitoes is to eliminate their breeding areas. Drain containers or any areas where water stands unnecessarily. Besides ponds, puddles, and boggy marshes, mosquitoes can breed in old tires, leaf-clogged gutters, empty cans and jars, birdbaths, ornamental ponds, wells, puddles from leaking plumbing, tree holes, and similar places. Any water that stands for a week or more during warm weather can be a source of adult mosquitoes. If you cannot drain the water, or if it is an ornamental pond that you do not want to drain, stock it with fish, or apply methoprene, an insect growth regulator that prevents the larvae from maturing into adults. You can also apply *Bacillus thuringiensis* var. *israelensis*, a bacterial insecticide that is specifically effective against mosquitoes.

Many areas of the country are served by mosquito control districts that are responsible for communitywide mosquito control. If you have undue numbers of mosquitoes, you can report it to the local agency, and the staff will assess the problem and determine how it can be solved. In many cases, they will do the control work at no charge.

First Aid
Wash the bite site with soap and water and apply benzocaine or cortisone to reduce any itching. An antihistamine taken orally reduces mild systemic symptoms. Consult a physician in the event of unusual symptoms.

SCORPIONS

The sting of most scorpions is not serious but can be quite painful. It is similar to a strong bee or wasp sting, producing local swelling, discoloration, and pain in the sting site. The sting of the sculptured or bark scorpion, found in Arizona, can be life-threatening, however, especially to young children and elderly people. A sting from this species produces a pins-and-needles sensation with no local swelling or discoloration. Numbness around the

sting site, rapidly spreading to the entire extremity, is common with these stings. The stung person may experience weakness, dizziness, and similar systemic symptoms.

Scorpions sting people only in self-defense. A person might be stung by inadvertently touching a scorpion when lifting an object from the ground or by putting his or her foot in a shoe into which a scorpion has climbed to hide during the day.

Scorpions occur throughout the United States but are most common in the southernmost areas of the country. You may encounter this pest in your yard or garden as well as in the wild. Indoors, they commonly infest attics and crawl spaces and may move into living areas if the attic temperature rises above 100° F.

New homes or ones less than three years old in scorpion-inhabited areas are especially likely to have scorpions indoors, probably because their territory was disturbed during construction. Homes located near dry riverbeds may have scorpions coming indoors during summer rains.

The Pest
The scorpion is a spider relative with eight legs, a pair of large pincers, and a six-segmented tail that ends with a bulbous segment and a prominent, curved stinger. The stinger contains the venom. A scorpion can move its tail very quickly in any direction and may sting its prey repeatedly.

Scorpions use their venom primarily to help subdue their prey. Their food usually consists of small spiders and soft-bodied insects. They grasp their prey with their pincers and sting by bringing their stingers forward directly over their heads.

Scorpions spend the day in burrows or under bark or objects on the ground. At night they emerge and wait for their prey. During the winter, or

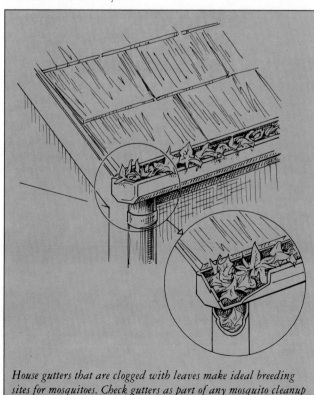

House gutters that are clogged with leaves make ideal breeding sites for mosquitoes. Check gutters as part of any mosquito cleanup program.

It's best to avoid all scorpions—not all are deadly, but their stings are painful.

any time the temperature drops below 77° F, they become inactive.

Female scorpions give birth to live young. Young scorpions ride on their mother's back until they go through their first molt (shedding of skin). They molt seven times, reaching maturity in about a year.

The most dangerous scorpion in the United States is the sculptured scorpion, or bark scorpion, as it is sometimes called. It is most common in Arizona but also occurs to some extent in New Mexico and adjacent areas of Texas, California, and Mexico. This scorpion is 2 to 3 inches long at maturity, straw-colored, and has blackish stripes on its back. The sculptured scorpion is very similar to other scorpions, and only an entomologist or other trained specialist can distinguish it from other species.

Unlike other scorpions in Arizona, the sculptured scorpion lives in trees and does not burrow. Groves of mesquite, cottonwood, and sycamore are common habitats for this scorpion.

Prevention and Control
If you live in an area where there are scorpions, always shake out or look inside your shoes, gloves, clothing, and sleeping bag before using them. Wear leather gloves whenever you move rocks, boards, and other possible scorpion hiding places.

Eliminate stacks of lumber, firewood, and rocks; compost piles; and similar items from your property. These can serve as hiding places for scorpions. Seal up cracks around windows and doors as well as other possible entry places.

If scorpions are common in the area, you may need to spray outdoors around the foundation to prevent them from coming indoors. Spray around the foundation, giving special attention to potential hiding places and entry sites, such as weep holes in brick. Use diazinon (Ortho Diazinon Insect Spray), Sevin® (carbaryl), or Baygon® (propoxur). Spraying at dusk, so that the insecticide is fresh when the scorpions come out at night, will maximize the effectiveness.

Control any scorpions indoors by applying an insecticide to baseboards, in cracks and crevices, under furniture, and other possible hiding places. Because scorpions frequent crawl spaces and attics, treat these areas as well. For crawl spaces, treat around the foundation and around all piers. Use a spray or dust of diazinon or Baygon® (propoxur) labeled for indoor use. Since scorpions are predators, eliminating cockroaches and other household pests makes a home less attractive to them.

First Aid
Even though most scorpion stings are no more dangerous than a bee sting, you should consider all stings dangerous and contact a physician immediately for treatment instructions.

SPIDERS

Most spiders cannot bite through human skin and cannot harm people. (See page 48 for information on household spiders.) A few species have stronger mouthparts and venom that produces either no symptoms or very mild ones. The black widow and brown recluse spiders are the most important exceptions— their mouthparts are strong enough to penetrate human skin, and their venom is potent. These dangerous spiders adapt well to living in and around human-made structures.

A black widow bite may either initially go unnoticed or produce a sharp pain followed by a dull, numbing pain. The pain peaks after one to three hours and persists for one to three days. Other effects range from mild to severe and include muscular spasms, rigid abdominal muscles, and tightness in the chest. Most of

small blister. In the next day or two the skin at the bite becomes discolored, and a scab develops by the end of the first week. When the scab comes off, the bite site sometimes develops into an open, ulcerating wound that can take many weeks to heal. The victim may also experience nausea, fever, chills, and other symptoms for up to a week after the bite.

Some tarantulas in tropical parts of the world can give a very painful bite, but the species occurring in North America generally have a mild bite that is painful for no more than a few minutes. Some tarantulas have hairs that can cause intense itching when they touch your skin.

These and all other spiders are generally shy and usually retreat from people. Most bites occur when the spider is squeezed or seriously provoked. This can happen when you lift an object and accidentally touch a spider or

Tarantulas usually hide in burrows and hunt only at night. If you see a wandering tarantula, it is probably an escaped pet or a male seeking a mate.

these bites leave no visible mark, but some leave two tiny, red puncture spots. Small children, the elderly, and people with health problems are the ones most affected by the bite of this spider. Approximately 5 percent of untreated black widow bites result in death. The bite is never fatal if a physician treats it promptly.

Most brown recluse bites go unnoticed at first. The initial symptoms can begin several hours later and include local pain, redness, and a

when the spider is in a bed sheet or article of clothing. Sometimes a spider will bite when you disturb its web, apparently thinking that the vibrations are caused by trapped prey. Some spiders will bite to defend their eggs or when they are cornered.

The Pest
When full grown, the female black widow has a rounded, glossy, black body about ½ inch long, with legs about an inch long. Two reddish marks on its underside may form a

Above: Viewed from above, female black widow spiders appear entirely black; the red hourglass marking is on the underside. Below: Black widows build tangled webs in dark corners. The round, gray object is an egg sac; a male black widow is in the web below and to the left of the egg sac.

distinct or irregular hourglass shape. The mature male and immature spiders of both sexes are lighter in color with cream markings on the tops of their abdomens and a yellowish to red hourglass mark on their undersides. Only mature and larger immature females can bite; males are harmless. Contrary to popular belief, a female black widow rarely eats the male after mating.

Black widow spiders occur throughout the United States. They build irregular, tangled cobwebs with strands that are thicker than those of most other spiders. These are usually in dark, secluded places near ground level. Common web sites include piles of wood or stones, animal burrows, garages, sheds, utility meter boxes, and campground outhouses. Most bites occur when a person inadvertently touches a black widow's web or puts his or her hand directly on the spider, often while picking up an object under which the spider is hiding.

The brown recluse spider is yellowish tan to dark brown with a characteristic violin-shaped mark behind its head. The mature female is about ¹/₂ inch long with legs that are about ³/₄ inch long; the male is slightly smaller. Immature brown recluse spiders have the same coloring as adults. Both males and females bite.

Brown recluse spiders make irregular webs without any definite pattern. They use their webs primarily as resting places during the day, rather than as means of catching insects. At night they rove about in search of prey. Most bites occur when a person rolls over on one in bed or puts on clothing or shoes that have not been worn for a long time. As their name implies, these spiders are reclusive and will retreat if they can. They spend most of their lives in out-of-the-way places. Indoors, they hide in boxes, corners, crevices, and old clothes. Outdoors, they may be under rocks, in lumber piles, and among rubbish.

Brown recluse spiders are most common in Arkansas,

Kansas, Missouri, Oklahoma, and surrounding states, but they have also been reported in many other areas. They spread to new areas primarily in shipping cartons and on household goods. Several closely related species are practically identical in appearance and habits, but their venom is not quite as potent. These occur in many other states, primarily in the southern half of the country and on the East Coast.

Tarantulas are large, hairy spiders up to 5 inches across. They live in dark cavities or burrows and hunt at night. These spiders take 10 years to reach maturity. Most tarantulas enter homes during the summer months and are males looking for mates. They live primarily in the Southwest.

Tarantulas are so gentle that many people keep them as pets. They are sluggish and rarely bite unless mishandled.

Prevention and Control

Control black widow and brown recluse spiders indoors by spraying their webs, along baseboards, and similar potential hiding places with Baygon® (propoxur—Ortho Ant, Roach & Spider Killer), Dursban® (chlorpyrifos—Ortho Home Pest Insect Control), or tetramethrin and sumithrin (Ortho Home & Garden Insect Killer Formula II) according to label directions. You can use one of these sprays or a flyswatter to kill individual spiders. A fogger is useful to clear a room of exposed spiders and their prey but will not penetrate into cracks and crevices and other out-of-the way places.

Spray areas outdoors that are infested with black widow or brown recluse spiders with Dursban® (chlorpyrifos—(Ortho-Klor® Indoor & Outdoor Insect Killer), diazinon (Ortho Diazinon Insect Spray), malathion, or Baygon® (propoxur). Pay particular attention to areas near the home where children play, woodpiles, and similar places.

In severely infested areas, use one of these products to treat around the outside foundation, in crawl spaces, under eaves, and around windows and other openings to control spiders.

If you live in an infested area, always wear gloves when picking up firewood or debris, shake out clothing and look inside shoes before dressing, and check towels and bedding before using them. Vacuum underneath furniture and in storage areas to remove spiders, webs, and egg sacs. Remove old boxes and papers, piles of building materials, trash, and other unwanted items from around the house and from basements and living areas. Be sure that all windows and crawl-space and

Brown recluse spiders rarely venture into the open, but if you get a close look you can identify one by the dark brown violin shape behind its head.

roof vents have screens to help prevent the spiders from entering. Control other insects that serve as food for spiders.

You can capture a stray tarantula in a box or sweep it up in a dustpan and empty it into a large grocery bag. They seldom bite, but it is a good idea to avoid sudden, quick movements and to wear gloves for additional protection.

First Aid

See a physician as quickly as possible if you suspect you have been bitten by a black widow, brown recluse, tarantula, or any spider that causes an unusual or severe reaction. Capture the spider if practical to aid in diagnosing and treating the bite.

TICKS

Ticks suck blood from a variety of warm-blooded animals, including people and their pets. The actual bite is not painful, and it usually goes unnoticed. Some tick bites leave a bright red spot. Sensitive people may experience some swelling and irritation. There may be a secondary infection if the mouthparts of the tick are broken off and left in the skin. Large numbers of ticks on a dog can suck enough blood to make the dog irritable and sluggish.

Ticks can transmit Rocky Mountain spotted fever, typhus, tularemia, tick paralysis, Lyme disease, and Colorado tick fever. The brown dog tick can also breed indoors and become abundant inside homes. Other species usually feed for shorter times and rarely infest homes.

The Pest
Ticks are oval or flattened and generally reddish brown to dark brown. They have no antennae, and the head, thorax, and abdomen are all fused together. Larval ticks may be only $1/40$ inch long. When engorged with blood,

Before engorging, the tick is a flattened brown or reddish brown creature. Pictured is an American dog tick.

the adults of some species can be up to $1/2$ inch long. Ticks may take several days to engorge themselves before they drop off their hosts. Like mites, tick nymphs have only three pairs of legs; older nymphs and adults have four pairs. Ticks in each stage of development need a blood

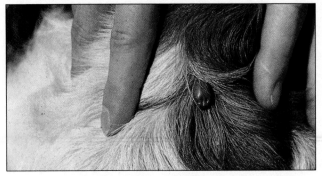

Dogs that roam in fields and woods should be examined regularly for ticks. Here a search has revealed an engorged tick.

meal before they can go on to the next stage. Many ticks can complete a life cycle in about three months in warm weather and with suitable hosts. They can often go for over a year without food.

Ticks are most abundant in brushy areas where vegetation is sufficient to support both large and small animals. They are more common on vegetation along animal and human trails than in areas where they are not as likely to encounter a passing host.

Of the many species of tick, brown dog ticks and American dog ticks are the ones most often encountered around homes. Brown dog ticks are a distinctive uniform reddish brown color. The adults are about $1/8$ inch long before feeding and up to $1/2$ inch long after they feed and become engorged with blood. These are the ticks usually encountered indoors.

Unlike other ticks, brown dog ticks in all stages will feed on dogs, so they can go through a complete life cycle indoors, in yards, and in kennels if a dog lives in these areas. Indoors, the females lay their eggs in cracks behind baseboards and ceiling boards, under carpeting, and in similar places. They usually attach to a dog's ears, the midline of the back, and between the toes. These ticks feed almost exclusively on dogs, seldom attacking other mammals and only rarely biting people. They are not found in woods unless a dog frequents the area.

Brown dog ticks occur in most states but develop larger populations in warmer regions. They probably can't survive the winter outdoors in areas colder than North Carolina.

American dog ticks are dark brown mottled with white. The adult females are $1/4$ inch long before feeding and up to $1/2$ inch long after feeding. Meadow mice are the most important hosts for the immature stages. The adults feed principally on dogs and large animals, but they will also feed on humans and a wide variety of wild animals. These ticks usually attach to the sides of a dog's face, the neck, and the shoulder region. They occur throughout most of the United States.

American dog ticks can infect people with Rocky Mountain spotted fever. This possibly deadly disease, characterized by a high fever and a rash, is rare but can occur throughout the country. Recently, most cases have been reported in Oklahoma, Ohio, Virginia, Maryland, and South Carolina.

Deer ticks are about $3/16$ inch long before engorgement. They feed on a variety of mammals and birds, including deer and dogs, and they also bite people. These ticks rarely infest homes but may be present in grassy or woodsy areas near homes built in wooded areas. Deer ticks can carry Lyme disease, a serious and newly recognized arthritic illness. It is initially characterized by a fever, a rash, and flulike symptoms. The disease affects dogs as well as people. Lyme disease has been found in 32 states but is most prevalent along the North Atlantic coast from Massachusetts to North Carolina, along the Pacific coast in northern California and Oregon, in eastern Texas and Louisiana, and in Minnesota and Wisconsin.

Prevention and Control
Control ticks around the outside of the house by spraying with diazinon (Ortho Diazinon Insect Spray), Dursban® (chlorpyrifos—Ortho Outdoor Ant, Flea & Cricket Spray or Ortho Flea-B-Gon® Flea & Tick Killer), Baygon® (propoxur—Ortho Ant, Roach & Spider Killer), Sevin® (carbaryl), or malathion according to label directions. Pay particular attention to weedy areas, lawns, and other areas where your pet walks or sleeps.

To help prevent ticks from becoming established around the outside of the house, keep lawns and weeds mowed, and remove piles of debris that might harbor rodents. Regularly examine your pets for ticks. Keep wild animals from nesting under the house by screening any openings they might use.

To control ticks inside the house, spray or dust with Dursban® (chlorpyrifos—Ortho Flea-B-Gon® Flea & Tick Killer or Ortho-Klor® Indoor & Outdoor Insect Killer), Baygon® (propoxur—Ortho Ant, Roach & Spider Killer), malathion, or diazinon according to label directions. Treat areas where your pet sleeps, as well as along baseboards, doorway and window frames, and similar areas. If your pet is allowed on upholstered furniture, remove the cushions and treat any cracks and seams, as well as the rug underneath. Whenever you treat fabric, use a light spray or dust, and test for staining before you treat visible portions. Treat the crawl space of the house if it is accessible to pets or other animals.

Remove ticks from pets using the same techniques you use for humans (see First Aid). Dust or shampoo the animal with one of the insecticides labeled for this purpose, such as Sevin® (carbaryl—Ortho Sevin® Garden Dust). A veterinarian can also treat animals for ticks. Clean or remove your pet's bedding material. If you are treating for ticks in or around the house, treat your pet on the same day.

To help avoid tick bites while in tick country, stay on wide paths and roads when possible, and tuck in your pant legs and shirttail. Use an insect repellent containing diethyltoluamide (deet) on exposed parts of your body and on your clothing. Examine your body once or twice a day for ticks. When choosing a campsite, drape a white towel or piece of flannel over the grasses and shrubs. If any ticks are present, you can see them easily on the cloth.

First Aid

Remove attached ticks as soon as possible. The longer a tick remains attached, the greater the chance that it can infect its host with a tick-borne disease. The tick may have buried its head in your skin, and so must be removed cautiously. If the head is left inside, it can cause an infection. Although applying a just-extinguished kitchen match to the tick or covering it with fingernail polish or petroleum jelly may cause it to back out, this is not the best method. Research has shown that ticks are best removed with a pair of flat-tipped tweezers. Grasp the tick as close to the skin surface as possible, and pull upward with a slow, steady motion. Do not jerk or twist, as this may break off the tick's mouthparts. Do not squeeze or crush the tick on bare skin because its body fluids may contain infective agents. Do not handle the tick with your bare hands; wear

rubber gloves if you need to touch it. After removing the tick, thoroughly disinfect your hands and the bite site with soap and water. Kill the tick with rubbing alcohol, and then flush it down the toilet.

See a physician if you are unable to remove a tick completely, if the wound becomes infected, or if you become ill or have unusual symptoms following a tick bite or after being in a tick-infested area. Health authorities are increasingly concerned about the spread of Lyme disease. Vacationers may contract the illness and be unaware of a tick bite, and their home-town physicians may not be familiar with the symptoms. Lyme disease is curable if treated early enough, however.

YELLOWJACKETS AND OTHER WASPS

Yellowjacket and wasp stings produce the same symptoms as a bee sting. Most victims suffer only mild pain for just a few minutes, but extremely sensitive people can suffer lethal or near-lethal anaphylactic shock unless treated immediately.

The terms "yellowjacket" and "hornet" can be confusing because some people use them interchangeably while others use them to designate specific species. A hornet is a certain type of yellowjacket but is larger than other yellowjackets. Their behavior is similar. The only true hornet in North America, the European or giant hornet, occurs primarily in the northeastern states.

Yellowjackets, hornets, and umbrella wasps are all social insects that live in colonies consisting of many workers and one queen. Other wasps live solitary lives in which each female lives separately and provides for its young itself. These wasps include mud daubers, tarantula hawks and other spider wasps, cicada killers, and many more.

Above: Although yellowjackets feed pest insects to their young, and so are somewhat helpful, they become pests themselves when they nest too near a home or attend a picnic.
Below: An underground yellowjacket nest represents a hazard for the unwary person who does not look where he or she walks.

Solitary wasps are somewhat less dangerous because they don't attack in groups, as social wasps do. Some solitary wasps have a very mild sting, whereas the sting of others is more potent than those of social wasps.

The Pests

Yellowjackets and hornets are the most temperamental and aggressive wasps. Their nests always have just one entrance, and the shape of the nest is somewhat like a football. They build their nests in abandoned rodent burrows and in cavities in homes and trees, as well as under eaves and branches of trees and shrubs. Yellowjackets are usually yellow and black and about 1/2 inch long.

Umbrella wasps look like yellowjackets and have just as venomous a sting. They are less likely to sting unless their nest is seriously disturbed. These wasps never build their nests underground. Favorite places are beneath eaves and tree limbs. The nest is attached by a single slender

stalk and is shaped like an inverted umbrella, with each cell exposed and facing down.

Mud daubers are solitary wasps that build conspicuous nests of mud or clay against the sides of houses, under eaves, and in similar locations. These wasps have long, narrow waists and are dark or metallic-colored. The females provision each cell of their nests with one or more paralyzed spiders, lay an egg on the spiders, and then seal up the cells. Mud daubers rarely defend their nests, and they don't have an especially painful sting.

Spider wasps are medium to large and are often metallic blue or black. The female hunts for spiders and paralyzes them with its stinger. It then prepares a cell in the ground, drags the spider into the cell, lays an egg on the spider, and seals up the cell. The tarantula hawk is a type of spider wasp that preys exclusively on tarantulas. It doesn't sting people unless captured with bare hands or pinched. The sting is very painful.

Although less aggressive than yellowjackets, umbrella wasps do sometimes sting if they are threatened.

Mud daubers rarely sting. When the cells of their mud nests are open, the young have hatched, and will not return.

Cicada killers are up to 1½ inches long and brown or black with yellow markings. They paralyze cicadas with their stingers and fly them back to cells in the ground. They don't sting people unless captured with bare hands or pinched. Their sting is not particularly painful.

Velvet ants look like hairy ants without wings but are actually wasps that parasitize the pupae of other solitary wasps. Some species roam about on sandy beaches. They have a painful sting when stepped on with a bare foot.

Colonies of yellowjackets and other social wasps die in the fall, and only the newly mated females survive over winter. These spend the winter in protected places such as under bark, shingles, and stones. In the spring they establish new nests under eaves or branches. The nest material is a thin, papery substance made of soft, often decaying wood or vegetable fibers and mixed with saliva to bond it together. When the first workers emerge, they take over caring for the young, and the queen only lays eggs. In late summer, the colony begins to produce new queens and males, and egg laying ceases.

Larval yellowjackets and wasps eat high-protein foods, such as other insects and bits of meat, that the adults bring to them. The adults eat primarily carbohydrates, such as nectar, fruit juices, soft drinks, and other sweet substances. By midsummer the colony produces fewer young and the adult population has grown very large. This is when these insects become a real nuisance, because the adults search for sweets around picnic tables and on ripe fruit. By early fall the adults begin dying, and the problem quickly subsides.

Prevention and Control

Since wasps are beneficial predators on other insects, do not eradicate them unless necessary. A wasp nest next to a doorway, pathway, children's play area, or similar location where it presents a hazard should be controlled. Use an aerosol wasp bomb that shoots a narrow stream of spray at least 10 to 20 feet. This allows you to stand away

Solitary ground-nesting wasps help control pest beetles, caterpillars, and other insects. The wasp larvae eat the paralyzed pests.

from the nest while spraying. These products contain Baygon® (propoxur—Ortho Hornet & Wasp Killer), resmethrin (Ortho Outdoor Insect Fogger), or any of a number of other insecticides.

Wear a hat, rubberized gloves, a thick, long-sleeved shirt, and long pants for extra protection when treating a wasp nest. Secure your pants and jacket cuffs tightly over your shoe tops and gloves. Treat the nest after dark or in the early morning when the nest is inactive and the wasps are inside. Calmly walk inside your home after treating the nest—do not run away. Wasps sometimes follow moving lights such as flashlights, so either don't use one or cover the light with red cellophane to make it invisible to the wasps. Repeat the treatment at the same time the next day if you see activity in the nest.

Use these same sprays to control yellowjacket nests in the ground. Follow the same precautions just described. Always approach ground nests quietly; vibrations from heavy footsteps may alert the yellowjackets.

To prevent wasps from coming to your yard and scavenging for food, be sure that your garbage cans have tight lids, and pick up and dispose of fallen, ripe fruit from fruit trees every day. Keep any food covered so that wasps don't discover it and communicate its location to other wasps.

In those situations where yellowjackets are continually flying into your yard, try to find and treat the nest directly. You can sometimes follow a yellowjacket back to its nest, but this often proves difficult. Often the nest is within 2,000 feet of where the wasps forage for food, but it can also be more than a mile away. In such cases, you can try trapping the workers with a commercial wasp trap containing an attractant (Ortho Yellow Jacket Trap).

Kill wasps that fly indoors with a flyswatter or household aerosol spray, or capture them in a jar or box to release outdoors. If many wasps come indoors, check to see where they enter and seal up the opening or screen over it.

You can decrease the chance of being stung by not wearing perfumes, hair spray, suntan lotion, cosmetics, and bright colors outdoors, especially from midsummer to early fall. If a bee or wasp flies into the car, remember that it won't attack unless you strike at it or sit on it or otherwise trap it against yourself. Keep calm, stop the car when it is safe to do so, and then gently shoo it out an open window or mash it with a wadded-up tissue.

First Aid

Treat yellowjacket and wasp stings as you would a bee sting (see page 78). The stinger does not detach from the insect, however, so scraping the skin to remove it is unnecessary. Because yellowjackets are scavengers, their sting is not as clean as a bee sting, and you should thoroughly wash the sting site with soap and water and apply an antiseptic.

INDEX

U.S. MEASURE AND METRIC MEASURE CONVERSION CHART

		Formulas for Exact Measures			Rounded Measures for Quick Reference		
	Symbol	When you know:	Multiply by:	To find:			
Mass (Weight)	oz	ounces	28.35	grams	1 oz		= 30 g
	lb	pounds	0.45	kilograms	4 oz		= 115 g
	g	grams	0.035	ounces	8 oz		= 225 g
	kg	kilograms	2.2	pounds	16 oz	= 1 lb	= 450 g
					32 oz	= 2 lb	= 900 g
					36 oz	= 2¼ lb	= 1,000g (1 kg)
Volume	tsp	teaspoons	5.0	milliliters	¼ tsp	= 1/24 oz	= 1 ml
	tbsp	tablespoons	15.0	milliliters	½ tsp	= 1/12 oz	= 2 ml
	fl oz	fluid ounces	29.57	milliliters	1 tsp	= 1/6 oz	= 5 ml
	c	cups	0.24	liters	1 tbsp	= ½ oz	= 15 ml
	pt	pints	0.47	liters	1 c	= 8 oz	= 250 ml
	qt	quarts	0.95	liters	2 c (1 pt)	= 16 oz	= 500 ml
	gal	gallons	3.785	liters	4 c (1 qt)	= 32 oz	= 1 liter
	ml	milliliters	0.034	fluid ounces	4 qt (1 gal)	= 128 oz	= 3¾ liter
Length	in.	inches	2.54	centimeters	⅜ in.	= 1 cm	
	ft	feet	30.48	centimeters	1 in.	= 2.5 cm	
	yd	yards	0.9144	meters	2 in.	= 5 cm	
	mi	miles	1.609	kilometers	2½ in.	= 6.5 cm	
	km	kilometers	0.621	miles	12 in. (1 ft)	= 30 cm	
	m	meters	1.094	yards	1 yd	= 90 cm	
	cm	centimeters	0.39	inches	100 ft	= 30 m	
					1 mi	= 1.6 km	
Temperature	°F	Fahrenheit	5/9 (after subtracting 32)	Celsius	32°F	= 0°C	
					68°F	= 20°C	
	°C	Celsius	9/5 (then add 32)	Fahrenheit	212°F	= 100°C	
Area	$in.^2$	square inches	6.452	square centimeters	1 $in.^2$	= 6.5 cm^2	
	ft^2	square feet	929.0	square centimeters	1 ft^2	= 930 cm^2	
	yd^2	square yards	8361.0	square centimeters	1 yd^2	= 8360 cm^2	
	a.	acres	0.4047	hectares	1 a.	= 4050 m^2	

Proof-of-Purchase
0-89721-144-8